Water Residuals and Biosolids: Effect of Co-Application on Soil Phosphorus

Awwa Research Foundation

About the Awwa Research Foundation

The Awwa Research Foundation (AwwaRF) is a member-supported, international, nonprofit organization that sponsors research to enable water utilities, public health agencies, and other professionals to provide safe and affordable drinking water to consumers.

The Foundation's mission is to advance the science of water to improve the quality of life. To achieve this mission, the Foundation sponsors studies on all aspects of drinking water, including supply and resources, treatment, monitoring and analysis, distribution, management, and health effects. Funding for research is provided primarily by subscription payments from approximately 1,000 utilities, consulting firms, and manufacturers in North America and abroad. Additional funding comes from collaborative partnerships with other national and international organizations, allowing for resources to be leveraged, expertise to be shared, and broad-based knowledge to be developed and disseminated. Government funding serves as a third source of research dollars.

From its headquarters in Denver, Colorado, the Foundation's staff directs and supports the efforts of more than 800 volunteers who serve on the board of trustees and various committees. These volunteers represent many facets of the water industry, and contribute their expertise to select and monitor research studies that benefit the entire drinking water community.

The results of research are disseminated through a number of channels, including reports, the Web site, conferences, and periodicals.

For subscribers, the Foundation serves as a cooperative program in which water suppliers unite to pool their resources. By applying Foundation research findings, these water suppliers can save substantial costs and stay on the leading edge of drinking water science and technology. Since its inception, AwwaRF has supplied the water community with more than $300 million in applied research.

More information about the Foundation and how to become a subscriber is available on the Web at **www.awwarf.org**.

Water Residuals and Biosolids: Effect of Co-Application on Soil Phosphorus

Prepared by:
Jim Ippolito, **Mary Stromberger**, **Ken Barbarick**, and **Robin Bayley**
Department of Soil and Crop Sciences
Colorado State University
Fort Collins, CO 80523-1170

Sponsored by:
Awwa Research Foundation
6666 West Quincy Avenue, Denver, CO 80235-3098

Published by:

Awwa Research Foundation

American Water Works Association

IWA Publishing

DISCLAIMER

This study was funded by the Awwa Research Foundation (AwwaRF). AwwaRF assumes no responsibility for the content of the research study reported in this publication or for the opinions or statements of fact expressed in the report. The mention of trade names for commercial products does not represent or imply the approval or endorsement of AwwaRF. This report is presented solely for informational purposes.

CONTENTS

TABLES

FIGURES

FOREWORD

The Awwa Research Foundation (AwwaRF) is a nonprofit corporation that is dedicated to the implementation of a research effort to help utilities respond to regulatory requirements and traditional high-priority concerns of the industry. The research agenda is developed through a process of consultation with subscribers and drinking water professionals. Under the umbrella of a Strategic Research Plan, the Research Advisory Council prioritizes the suggested projects based upon current and future needs, applicability, and past work; the recommendations are forwarded to the Board of Trustees for final selection. The foundation also sponsors research projects through an unsolicited proposal process; the Collaborative Research, Research Applications, and Tailored Collaboration programs; and various joint research efforts with organizations such as the U.S. Environmental Protection Agency, the U.S. Bureau of Reclamation, and the Association of California Water Agencies.

This publication is a result of one of these sponsored studies, and it is hoped that its findings will be applied in communities throughout the world. The following report serves not only as a means of communicating the results of the water industry's centralized research program but also as a tool to enlist the further support of the nonmember utilities and individuals.

Projects are managed closely from their inception to the final report by the foundation's staff and large cadre of volunteers who willingly contribute their time and expertise. The foundation serves a planning and management function and awards contracts to other institutions such as water utilities, universities, and engineering firms. The funding for this research effort comes primarily from the Subscription Program, through which water utilities subscribe to the research program and make an annual payment proportionate to the volume of water they deliver and consultants and manufacturers subscribe based on their annual billings. The program offers a cost-effective and fair method for funding research in the public interest.

A broad spectrum of water supply issues is addressed by the foundation's research agenda: resources, treatment and operations, distribution and storage, water quality and analysis, toxicology, economics, and management. The ultimate purpose of the coordinated effort is to assist water suppliers to provide the highest possible quality of water economically and reliably. The true benefits are realized when the results are implemented at the utility level. The foundation's trustees are pleased to offer this publication as a contribution toward that end.

Walter J. Bishop
Chair, Board of Trustees
Awwa Research Foundation

Robert C. Renner, P.E.
Executive Director
Awwa Research Foundation

ACKNOWLEDGMENTS

The author of this report is indebted to the following water utilities and individuals for their cooperation and participation in this project:

City of Fort Collins Utilities – Wastewater, Fort Collins, CO

In addition, the advice and guidance of the Project Advisory Committee, including Jim DeWolfe of McGuire Malcolm Pirnie, Amit Pramanik of the Water Environment Research Foundation, and Dr. Chris Saint of the Cooperative Research Centre for Water Quality and Treatment (Australia), and the help of the AwwaRF project officer, India Williams, were much appreciated.

The author wishes to acknowledge the research assistance of Robin Bayley and the guidance of her MS committee, Ken Barbarick, Mark Paschke, and Mary Stromberger.

EXECUTIVE SUMMARY

Co-application of biosolids and water treatment residuals (WTR) has not been extensively studied. Biosolids inherently contain excessive phosphorus (P), and when repeatedly applied may contribute to soil P saturation, leading to off-site movement and eutrophication of adjacent surface water bodies. Alum-based [$Al_2(SO_4)_3 \cdot 14H_2O$] WTR are a by-product from the municipal drinking water treatment process, contain high levels of amorphous Al, and subsequently may have a large P sorption capacity. When WTR is applied alone to soils there is a concern that soil-borne P will be sorbed and reduce plant available P. Co-application of WTR with biosolids may benefit municipalities by sorbing excess biosolids-borne P, reducing the likelihood of waterway eutrophication, and in terms of cost saving by landfill avoidance. In the future, if municipalities are required to apply biosolids based on plant P requirements, not N requirements, co-application with WTR could be considered a best management practice (BMP).

RESEARCH OBJECTIVES

The Awwa Research Foundation (AwwaRF) funded this study to determine both the long-lasting effects of a single WTR-biosolids co-application (applied in 1991) and the short-term impacts of a repeated WTR-biosolids co-application (applied in 2002) on native rangeland soil phosphorus dynamics. This report presents the results of the study in terms of time of application (1991 or 2002) and changes between sampling periods, with field soil sampling occurring in October 2003 and October 2004. The key objectives of the study were to:

1. Quantify changes in inorganic soil P associated with a single or repeated WTR-biosolids co-application and identify which inorganic P phase dominates using a sequential extraction technique.
2. Quantify changes in organic soil P associated with a single or repeated WTR-biosolids co-application and identify which organic P phase dominates using a sequential extraction technique.
3. Identify shifts between organic and inorganic P fractions using P enzyme assays and determine organic P mineralization rates.

APPROACH

The city of Fort Collins, CO owns the 40,000 ha Meadow Springs Ranch and utilizes it for beneficial land application of biosolids. In August 1991, treatments were established on-site to assess short-term impacts of a single co-application of municipal biosolids and alum-based WTR on rangeland soil and vegetation. Test plots were 15 × 15 m with treatments consisting of three different WTR rates co-applied with a single biosolids rate (0, 10, and 21 Mg WTR ha^{-1} coapplied with 10 Mg biosolids ha^{-1}). The biosolids were applied with a side-discharge manure spreader, WTR by hand, and all treatments were replicated four times in a randomized complete block design.

In October 2002, using volunteers from Colorado State University, the city of Fort Collins wastewater utilities department, and the US Environmental Protection Agency, the same treatments at the same rates were reapplied, using a rear discharge spreader, to one-half of each 15 ×

15 m treatment plot. The study site at present contains a total of 3 WTR-biosolids treatments replicated 4 times in a randomized split-plot with time design for a total of 24 plots.

In October 2003 and October 2004, soil samples were collected from the 0-5-cm depth within each subplot using a mechanical hydraulic soil probe. Ten soil cores per subplot were obtained and composited for a total of 24 samples. Samples were placed in a cooler and transported to Colorado State University facility for subsequent analyses.

Inorganic Soil Analyses

Soil samples were air-dried and then ground to pass a 2-mm sieve. Inorganic soil analyses determination included pH, EC (salt content), total carbon (C) and nitrogen (N), and inorganic C.

Inorganic Soil P Analyses

A five-step inorganic P fractionation procedure determined soluble and loosely bound P, Al-bound P, Fe-bound P, reductant-soluble P (i.e. occluded by Fe coatings), and Ca-bound P. Soils were also analyzed for oxalate extractable P, Al, and Fe, and the soil phosphorus saturation index (PSI) was then calculated based on:

$$PSI = [P_{ox}] / [Al_{ox} + Fe_{ox}]$$

Where P_{ox}, Al_{ox}, and Fe_{ox} are oxalate extractable P, Al, and Fe.

In addition, soils were analyzed using several P test methods presented in the Colorado P Test Risk Index, including Mehlich 3-P, Bray 1-P, and Olsen's P.

Organic Soil P Analyses

A six-step organic P fractionation procedure determined labile inorganic P, total labile P, labile organic P via the difference between total labile P and labile inorganic P, microbial biomass P, moderately labile P (humic and fulvic acid associated P), and highly resistant, non-labile P.

Organic to Inorganic Transformations/Analyses

Mineralization of organic P to inorganic P was determined by assaying the following soil enzyme activities: phosphomonoesterase; phosphodiesterase; pyrophosphatase; and phytase.

CONCLUSIONS

Inorganic Soil Analyses

1. Co-application of WTR and biosolids increased soil organic C and N content. However, this increase was related solely to biosolids treatment, indicating that WTR did not contribute to the C or N increase.
2. Co-application did not affect soil pH.
3. The repeated co-application created an increase in soluble salts within the upper soil profile. Given time, a decrease in salt content was noted and related to leaching below

the 0-5-cm sampling depth. The increased salt content was most likely due to the repeated biosolids application, not WTR, used in this study, because biosolids and WTR applied in 2002 had EC values of 20.2 and 1.8 dS m^{-1}, respectively.

Inorganic Soil P Analyses

1. Over time, decreasing amounts of P were bound to the inorganic soil fraction as determined by a sequential extraction scheme. The inorganic soil fractionation scheme measured crystalline Al-bound P and not non-crystalline Al-bound P, which is indicative of WTR.
2. The soluble and loosely-bound P was the most dynamic of all inorganic fractions, receiving P from the organic biomass fraction, contributing to the amorphous Al phase (i.e. WTR phase), and also contributing to plant uptake.
3. The Fe-bound, Al-bound, and occluded P fractions are fairly stable over the long-term.
4. The Ca-bound P fraction was present even given the acidic pH of the system (5.2 to 5.9). This fraction shows signs of dissolution due to weathering, and appeared to be releasing P to the Fe-bound and amorphous Al fraction.
5. Oxalate extractable P increased between 2003 and 2004, and may be due to increased precipitation causing an increase in WTR weathering, thus exposing greater surface area for P sorption to occur.
6. The phosphorus saturation index (PSI), as determined by oxalate-extractable Al, Fe, and P, showed that soils which received biosolids only had significantly greater PSI values than soils which received WTR. Both the single and repeated co-applications of 10 Mg biosolids ha^{-1} + 21 Mg WTR ha^{-1} led to PSI reductions equal to or below background soil.
7. Even after a single or repeated co-application, the Olsen, Mehlich-III, and Bray-I extractable P concentrations were still considered a high or very high risk for off-site P movement. Concomitant research has shown that these three tests provide a poor measure of P content under WTR land application; a water extraction test is the best method for determining P content under WTR usage.

Organic Soil P Analyses

1. The organic fraction contained the greatest amount of P as compared to the inorganic and amorphous Al-fraction (i.e. WTR fraction).
2. The organic and inorganic fractions increased and decreased, respectively, from 2003 to 2004, indicating a net immobilization effect. Increases in organic P were likely due to either biosolids addition or increased plant matter inputs resulting from increased precipitation in 2004, or both.
3. Labile organic P increased in repeated applications versus single applications, most likely due to the recent addition of biosolids adding this fraction to the system. Labile organic P decreased with increasing WTR rate.
4. Biomass P decreased from 2003 to 2004, while all other fractions increased from 2003 to 2004. The decrease over the sampling period could have been due to P loss to both the inorganic fraction as well as the amorphous, non-crystalline fraction. The changes

in the biomass P fraction over the sampling period were only correlated to the phosphodiesterase enzyme.

5. The single and repeated co-application moderately labile fractions were greater than the control but similar to biosolids addition alone, suggesting biosolids addition under co-application conditions contributed to the moderately labile organic P fraction.

Organic to Inorganic Transformations/Analyses

1. Enzyme assays appeared to be unreliable as accurate methods for determining shifts between the organic to inorganic pools (i.e. mineralization). However, some interesting results were discovered:
   a. Phosphomonoesterase and phytase activity increased dramatically with WTR addition. The WTR addition may have created a P deficiency-like system, causing plants to respond and release both phosphomonoesterase and phytase into the soil environment. Previous studies have shown that plants can exude these enzymes under P deficient conditions. The WTR treatments in this study did not produce observable plant P deficiency symptoms, but it is possible that the WTR caused a false reaction and caused plants to respond accordingly.
   b. Phosphodiesterase activity was greater in soils receiving biosolids only or untreated control soils than observed following WTR additions.
   c. Pyrophosphatase activity decreased with increasing WTR rate in 2004. Single co-application treatments resulted in greater enzyme activity levels when compared to repeated co-applications of WTR and biosolids. This could be due to a lack-of-need to degrade polyphosphates in the first few years following biosolids application because biosolids supplied necessary P.
   d. The phytase enzyme assay was correlated to humic acid P in both 2003 and 2004. It was possible that the humic acid fraction had a consistent amount of substrate over time. The phytase enzyme assay showed the most promise for use as an organic P fractionation predictor for future research.

RECOMMENDATIONS

1. This study was successful in determining dominant P pools and P movement between these pools over time. There were no indications that the biosolids-WTR co-application treatments caused P overloading or P deficiencies at the application rates used (5, 10, or 21 Mg WTR ha^{-1} co-applied with 5 Mg biosolids ha^{-1}), two concerns of biosolids or WTR application separately. Phosphorus sorption to amorphous Al as a result of WTR treatments will provide long-term P sorption stability and should reduce runoff risk. In regards to P dynamics, biosolids-WTR co-application is a viable option for municipal nutrient recycling and reduction of re-usable materials entering landfills.
2. In the future, if municipalities are required to apply biosolids based on plant P requirements, not N requirements, co-application with WTR could be considered a best management practice. The co-application of 10 Mg biosolids ha^{-1} + 21 Mg WTR ha^{-1} led to phosphorus saturation index (PSI) reductions equal to or below background soil. If municipalities were required to co-apply these materials as to not drastically alter a

system's ability to adsorb P, a 2:1 ratio of WTR to biosolids appears to be the optimal co-application ratio for the materials tested.

3. In the US, WTR are not generated to the extent of biosolids. Thus, co-application as a best management tool should be used in areas where P runoff or leaching risk is high. These areas would include soils adjacent to water bodies and soils with a high risk for P leaching.

FUTURE RESEARCH

This project has led to several interesting questions that should be answered. Suggested studies should research the effect of: runoff in terms of soil P release, causing particles to enter water bodies of varying pH, potentially affecting waterway eutrophication; a single or repeated co-application on plant species biomass, number, and overall ecosystem health; a single or repeated co-application on sorption of other anions, such as arsenate/arsenite, selenate/selenite, and molybdate. As far as can be determined, land co-application effects on sediment P release, ecosystem sustainability, and sorption of other anions has not been investigated.

CHAPTER 1
INTRODUCTION AND BACKGROUND

Water treatment residuals (WTR) and biosolids are both by-products from municipal treatment processes. Water treatment residuals, also referred to as alum sludge, is often considered a waste product from drinking water treatment facilities. Alum [$Al_2(SO_4)_3 \cdot 14H_2O$] is the main component used in the treatment process to destabilize colloids for flocculation and water clarification. Biosolids are a by-product of wastewater treatment. Both products have been studied separately for their effects and benefits for land application as an alternative method of beneficial reuse.

Some beneficial effects on soil have been attributed to WTR, including additions of organic carbon, improved soil structure, and improved water holding capacity. Elliott et al. (1990) reported a range of 0.85 to 6.49% total organic carbon in WTR, with an average of 2.99%. Elliott et al. (1990) also stated "the organic carbon is stable and resistant to degradation, similar to the organic material in soils." Rengasamy, Oades, and Hancock (1980) observed increased soil aggregation and water holding capacity with the addition of wet alum sludge. A study by Bugbee and Frink (1985) investigated different potting media mixes, including additions of alum. A treatment using equal parts soil, perlite and peat mixed with 67% alum showed significant improvements in aeration and increased water availability.

Even given the aforementioned benefits of WTR usage, land application still has resulted in mixed reviews. Cornell and Westerhoff (1981) state "attempts to use coagulation sludges as soil conditioners or stabilizers have had little success." The product is relatively inert, but there is concern over excessive addition of trace metals to the soil environment and the possibility of Al toxicity to plants (CDPHE 1994, 1996). Phytotoxicity as a result of Al is a concern in many agricultural situations. Aluminum is abundant in all soils and according to Elliott et al. (1990) "phytotoxicity is directly related to conditions controlling Al solubility and not the total amount in the soil. If the soil pH is maintained in the 6 – 6.5 range by liming or by sludge addition, Al toxicity does not pose a problem."

While metal toxicity is a consideration, the overwhelming concern with WTR land application is the amorphous metal oxide concentration having the capacity to adsorb tremendous amounts of P. When applied alone, WTR could significantly reduce plant available P via surface adsorption. Bugbee and Frink (1985) reported that lettuce (*Lactuca sativa,* cv. iceberg) grown in alum-amended soils took on a purple hue, characteristic of P deficiency. Leaf tissue and soil tests confirmed that P was less available to plants grown in that media. One exception existed in the media that contained equal parts soil, perlite, and peat with 67% alum where it was believed that the improvements in aeration and moisture holding capacity overshadowed the effects of P deficiency.

Heil and Barbarick (1989) performed a greenhouse study to observe the effects of WTR additions on sorghum-sudangrass (Sorghum bicolor L.). Phosphorus deficiencies were observed at the highest level of WTR application (25 g kg^{-1}). Deficiency effects were diminished with a doubling of P fertilizer, resulting in greater dry plant matter production, further illustrating the adsorption effect by WTR.

Similar results were reached in a study by Lucas et al. (1994). Fescue (*Festuca arundinaceae* Schreb.) was grown in soil amended with variable rates of WTR (0 to 4 percent by weight), P (50 and 100 mg P kg^{-1} soil) and presence or absence of lime. Linear yield decreases were observed for both P levels while lime was not a significant factor. Although yields decreased as WTR rate increased in both cases, yields were higher for the higher P application rate, supporting the adsorption effect by WTR. In support, Ippolito, Barbarick, and Redente (2002) state "high application rates of WTR should be avoided due to its adverse effect on P availability to plants, unless a supplemental P source is supplied."

While results from the previous studies indicated that WTR can decrease plant available P, they also illustrate that a supplemental P source can negate the adsorptive effects. These findings have provided a basis for further studies in which WTR is applied in a high nutrient concentrated area or combined with another by-product to create a more balanced product. Novak and Watts (2004) studied the use of WTR in coastal sandy soils where long-term application of animal manure created an environment with excessive P concentrations. They found that WTR significantly increased the adsorptive capacity of sandy soils.

The presence of P in sandy coastal areas and other areas near water is a major concern for surface water quality. According to a review by Correll (1998) P is the most common cause of eutrophication in fresh water lakes, reservoirs, streams, and headwaters of estuarine systems. High concentrations of soluble P in soils that are in close proximity to surface waters create a high risk situation for P movement into those waters. The addition of WTR to these areas could effectively reduce P concentrations to low-risk levels.

Makris et al. (2004) studied the long-term stability of Fe and Al-based WTR-sorbed P. They found that while WTR amorphous Al has a large surface area, surface area alone does not account for all sorbed P. It is possible for P to reside in micropores that are not accounted for when measuring surface area of WTR, and that micropore P is extremely resistant to desorption leading to long-term stability. Ippolito et al. (2003) found similar results, with WTR more strongly binding P over time since the initial P application. They also found that WTR sorbed 1.25% P, most likely due to increased microporosity and thus increased WTR surface area. Makris et al. (2004) noted that WTR adsorbed nearly 10,000 mg P kg^{-1} within ten days and lost only 0.2% to 0.8% after 80 days, asserting that micropore-bound P will remain sorbed in circumneutral pH aqueous media. Both the Ippolito et al. (2003) and Makris et al. (2004) studies concluded that WTR provides long-term stability of sorbed P.

Dayton and Basta (2005) found that WTR sorptive capacity can be estimated to optimize WTR use with materials containing excess P. The study found a strong correlation between measured Al oxide and P sorption maximum in WTR. Accurately determining WTR P sorption maximum is essential to effectively use WTR for reducing soluble P in agricultural runoff water, reducing P solubility in agricultural soils or organic waste materials (biosolids, manure), and to calibrate WTR application (Dayton and Basta, 2005).

Co-application of WTR with biosolids is a relatively new concept; however, beneficial land application of biosolids alone is a widely accepted practice. According to an EPA publication entitled "Biosolids Applied to Land: Advancing Standards and Practices," the 1992 prohibition of ocean disposal promoted the increase in land application of biosolids. An estimated 7 million metric tons of biosolids were created in 2005. Approximately 66% of those biosolids will be used for land application and other beneficial applications (USEPA, 1999). Land application not only reduces the amount of space necessary for landfill disposal but also allows for economic nutrient recycling.

In a field study by Barbarick et al. (1992), biosolids and N fertilizer were applied to dryland winter wheat. The study compared grain yields, grain elemental concentration, soil nutrient concentrations at variable depths and economic income. Study results found that an agronomic application of 6.7 Mg biosolids ha^{-1}, added every other year over an eight year period, provided an economic benefit over commercial N fertilizer in many years. The biosolids treatments led to higher grain production and elevated soil nutrient concentrations. No grain or soil nutrient concentrations were considered hazardous to domestic livestock or humans.

Although P was not a direct focus of Barbarick et al. (1992), the study indicated that P levels rose steadily over time and had become a risk. Ippolito and Barbarick (2006) further analyzed the soil and found that soil P levels were considered high-risk according to the Colorado Phosphorus Index Risk Assessment (Sharkoff et al., 2003). They added 0 to 10 g of WTR to 10 g of soil and then extracted P using five different methods. They found that the Bray-I and Mehlich-III extraction methods significantly decreased extractable P, reducing the soil to a low risk category according to the Colorado Phosphorus Index Risk Assessment. The authors also found significant decreases in water-extractable P with increasing WTR application, and suggested using a water extract to predict P availability in WTRs amended soils. This study provided compelling evidence for the future investigation of co-application of biosolids and WTR.

The Colorado Department of Public Health and Environment has set forth guidelines for beneficial co-application based on the Colorado Biosolids Regulations, 4.9.0 (CDPHE 2004). The regulations state that co-application is not an option for root crops or low growing fruits and vegetables meant for direct human consumption. There are also provisions for groundwater, soil and plant tissue testing for a variety of parameters including total P, Al, and several heavy metals in areas that have received biosolids/WTR co-application.

Two recent studies examined the direct effects of co-application of biosolids and WTR. Harris-Pierce, Barbarick, and Redente (1993) studied the effects of co-application on aboveground plant biomass of four shortgrass prairie species. Blue grama (*Bouteloua gracilis* H.B.K. Lag), western wheatgrass (*Agropyron smithii*), buffalograss (*Buchloe dactyloides*), and fringed sage (*Artemisia frigida*) were observed in plots treated with variable WTR rates ranging from 5 to 21 Mg ha^{-1} and a constant rate of 11 Mg ha^{-1} biosolids. The study found no significant differences in biomass or tissue concentrations in any plant species.

A similar study was conducted by Ippolito, Barbarick, and Redente (2002) to examine and quantify the effects of different combinations of WTR and biosolids on western wheatgrass and blue grama. Grasses were grown in pots containing variable rates of both WTR and biosolids (0 to 10 g kg^{-1}). Results indicated that WTR reduced the available P to both grass species, although neither species showed any P deficiency or reduction in dry matter yields.

Plants absorb P in the form of orthophosphate (i.e. soluble P; see Figure 3.23), with some orthophosphate being supplied by the mineralization of soil organic matter. Thus, plant availability of soil organic P may be characterized by estimating the activity of indigenous phosphatase enzymes which convert organic P to inorganic P. According to Tabatabai (1994), the general name phosphatase has been used to describe a broad group of enzymes that catalyze the hydrolysis of both esters and anhydrides. Thus, the mineralization of organic to inorganic P could be estimated via the assay of phosphomonoesterase, phosphodiesterase, pyrophosphatase, and phytase enzymes.

Phosphomonoesterase is studied as either acid or alkaline, based on the soil pH conditions. This study employed the acid phosphatase due to the neutral to slightly acidic nature of the soils involved. Acid phosphatase catalyzes a reaction which cleaves orthophosphate from the organic monoester. According to Miller et al. (2001), some plants can secrete the enzyme under P-deficient conditions and the acid phosphatases hydrolyze monoester soil organic P at low pH, thereby increasing orthophosphate availability. In this case an increase in enzyme activity leading to an increase in available P should be observed. When orthophosphate has reached sufficient levels, a decline in monoesterase activity should be observed because the enzyme itself is inhibited by orthophosphate.

Phosphodiesterase and pyrophosphatase have not been extensively studied or researched. According to Tabatabai (1994), phosphodiesterase enzyme activity has been detected in plants, animals, and microorganisms; it is responsible for degrading nucleic acids. Phosphodiesterase is known to be inhibited by several factors, including orthophosphate. There does not appear to be any correlation between pH and optimum activity of phosphodiesterase or pyrophosphatase. Tabatabai (1994) states that pyrophosphatase is of interest because its substrate, pyrophosphate, is used as fertilizer. The enzyme is found naturally in soils and aids in fertilizer effectiveness by catalyzing the hydrolysis of pyrophosphate to orthophosphate.

Measuring phytase enzyme activity may also be of interest because phytic acid is the most abundant organic P compound in soil (Anderson, 1980). The phytase enzyme is responsible for cleaving orthophosphate from phytic acid in plant material (Applegate and Angel, 2004). The enzyme is abundant in nature but is most known for its use in commercial livestock feed. Almost half of the P in some feed materials is associated with phytin. Monogasteric animals, such as swine and poultry, lack the necessary phytase enzyme to take advantage of the P. Phytase enzyme added to feed allows for a greater P availability to the animal, reducing the need for additional P sources in feed products.

OBJECTIVES

The study objectives were to understand both the long-term effects of a single co-application of biosolids and WTR and the short-term impacts of a repeated co-application on soil inorganic and organic P dynamics. We utilized sequential extraction techniques to determine the dominant inorganic and organic P pools, and equilibrium shifts between these pools over time using P enzyme assays. Transformations between organic and inorganic pools were determined by assaying the phosphatase enzyme group and the phytase enzyme. Our hypothesis was that single and repeated co-applications will show similar P fractionation patterns and that P bound in the amorphous Al fraction will dominate all inorganic fractions. Based on previous findings by Ippolito et al. (2003) and Makris et al. (2005) we would expect the amorphous fraction of many Al-based WTR to dominate all soil P fractions, sorbing greater than 50% of soil P due to the inherently large surface area of this amorphous compound.

CHAPTER 2
MATERIALS AND METHODS

The city of Fort Collins, CO owns the 40,000 ha Meadow Springs Ranch and utilizes it for beneficial biosolids land application. In August 1991, treatments were established on-site to assess short-term impacts of a single co-application of municipal biosolids and alum-based WTR on rangeland soil and vegetation. Test plots were 15 × 15 m with treatments consisting of three different WTR rates (5, 10, and 21 Mg ha^{-1}) co-applied with a single biosolids rate (10 Mg ha^{-1}). The biosolids were surface applied with a side-discharge manure spreader, WTR by hand, and all treatments were replicated four times in a randomized complete block design.

In October 2002 the original plots were divided in half. One-half received a second application using the same treatments at the original rates. This operation created a randomized split-plot with time design with a total of 24 plots.

BIOSOLIDS AND WATER TREATMENT RESIDUALS ANALYSES

Biosolids and WTR used for this study were obtained from the city of Fort Collins, CO wastewater and drinking water treatment facilities, respectively. Fort Collins, CO biosolids are generated using anaerobic digestion (City of Fort Collins, 2006). The Fort Collins, CO water treatment facility uses source water from both reservoir and river fed systems. Raw water is treated with alum to enhance coagulation, flocculation, and sedimentation of particles (City of Fort Collins, 2004). Biosolids and WTR elemental composition was determined by $HClO_4$-HNO_3-HF-HCl digestion (Table 2.1; Soltanpour et al., 1996) followed by elemental analysis using inductively coupled plasma atomic emission spectrometry (ICP-AES). Biosolids NO_3-N and NH_4-N were determined following methods outlined by Mulvaney (1996), and pH (Thomas, 1996) and electrical conductivity (EC) using a saturated paste extract (Rhoades, 1996).

Climate data from both sampling years indicated that 2003 was a drier than 2004 (NOAA, 2003–2004). The majority of the 2003 precipitation came early in the year followed by a dry summer and fall, leading up to the October sampling. Precipitation in 2004 was distributed throughout the summer with slightly over 7.6 cm in the two months leading up to the October sampling. The research area receives 33 to 38 cm of mean annual precipitation (NRCS, 1980).

The soil at the research site is classified as an Altvan loam, 0-3% slopes. The Altvan series consists of deep, well drained soils that formed in mixed alluvial deposits (NRCS, 1980). Soil samples were collected in October 2003 and 2004. Ten soil cores from the 0-5-cm depth were obtained from each subplot using a mechanical hydraulic soil probe. The cores were mixed to form a composite soil sample. In 2004 soil samples were collected from an adjacent study to gather control information. Control samples consisted of four plots that received no biosolids or WTR, four plots that received 10 Mg ha^{-1} biosolids in 1991, and four plots that received 10 Mg ha^{-1} biosolids in 1991 and again in 2002. Soil sampling and compositing were performed identical to the co-applied plots. All soils samples were placed in a cooler and transported to Colorado State University for analysis.

Table 2.1
The 1991 and 2002 Fort Collins, CO biosolids and WTR analysis

Property	1991 biosolids	2002 biosolids	1991 WTR	2002 WTR
K (mg kg^{-1})	1896	420	4178	1785
P (mg kg^{-1})	16141	11350	550	545
Fe (mg kg^{-1})	4948	19050	19500	14576
Cu (mg kg^{-1})	547	162	44	36
Zn (mg kg^{-1})	772	254	30	33
Ni (mg kg^{-1})	19	5	10	6
Mo (mg kg^{-1})	16	1.9	1.4	0.4
Cd (mg kg^{-1})	5	0.6	0.1	0.1
Cr (mg kg^{-1})	40	6	17	8
Pb (mg kg^{-1})	119	7	2	<0.05
Ca (mg kg^{-1})	28361	ND*	3438	12474
Al (mg kg^{-1})	8618	12650	63300	59016
Organic N (mg kg^{-1})	41161	41750	3885	3485
NO_3-N (mg kg^{-1})	98	3	64	118
NH_4-N (mg kg^{-1})	3640	5442	51	9
pH	7.3	7.3	6.8	7.1
EC (dS m^{-1})	5.0	20.2	0.5	1.8

* Not Determined.

GENERAL SOIL ANALYSES

Soil samples were air dried and passed through 2-mm sieve before analysis. For inorganic soil analysis, pH (Thomas, 1996) and EC (Rhoades, 1996) were measured using a saturated paste extract, total C and N were determined using LECO-1000 (LECO Corp., St. Joseph, MI) analysis (Nelson and Sommers, 1996), and inorganic C was measured using the modified pressure-calcimeter method (Sherrod et al., 2002).

SOIL TEST P

Soils were also analyzed using three P test methods presented in the Colorado P Risk Index (Sharkoff, Waskom, and Davis, 2003). Briefly, the Colorado P Risk Index assessment considers risk associated with runoff class (includes soil-permeability class and slope), soil-test P concentrations, P application rates, P application method, and best-management conservation practices (i.e., residue management). Mehlich-III (Mehlich, 1984), Bray-I (Kuo, 1996), and Olsen's P (Kuo, 1996) extractions were used to determine extractable P concentrations. Solution P was colorimetrically determined as described below. Mehlich-III and Bray-I extractants are typically used on acidic soils such as our site. Olsen's is used for basic soils. However, we utilized the Olsen's to check for biosolids-borne Ca and Mg impact on soil P.

INORGANIC SOIL P FRACTIONATION

Inorganic and organic P fractionation was performed according to methods outlined by Kuo (1996). A five step inorganic fractionation separated soluble and loosely bound, Al-bound, Fe-bound, reductant soluble, and Ca-bound P. Extracts were filtered through a 0.2 μm membrane and analyzed colorimetrically following a modified ascorbic acid procedure with a 0.04 mg L^{-1} detection limit (Rodriguez, Self, and Soltanpour, 1994) and a spectrophotometer set at 882 nm. Due to insufficient color development of the reductant soluble fraction, the original method was modified based on research by Weaver (1974). A 2.5 mL aliquot was transferred to a 25 mL volumetric flask. Then, 1.5 mL of 5% ammonium molybdate solution, 15 mL of deionized water (DI), and 2.5 mL of color developing reagent (Rodriguez, Self, and Soltanpour, 1994) were added, and the solution brought to a final volume using DI. The final solution was allowed to stand for 30 minutes prior to P analysis.

OXALATE EXTRACTABLE SOIL FRACTION AND PSI

Oxalate extractable P, Al, and Fe (Loeppert and Inskeep, 1996) were measured using ICP-AES. The oxalate extractable P, Al, and Fe concentrations were used to determine the soil Phosphorus Saturation Index (PSI):

$$PSI = [P] / ([Al] + [Fe]) \qquad (2.1)$$

Where: [ ] = concentration in mmol kg^{-1}

The PSI has a scale from 0 to 1, with 0 indicating no P saturation and 1 indicating 100% P saturation. Total soil P was measured according to Kuo (1996), and analyzed using ICP-AES.

ORGANIC SOIL P FRACTIONATION

The organic sequential fractionation procedure separated labile organic, microbial biomass, moderately labile, fulvic acid, humic acid, and non-labile associated P. Phosphorus concentrations were determined colorimetrically as previously described. Microbial biomass P was calculated as total labile P + $CHCl_3$-lysed microbial cells minus total labile P. Labile organic P was calculated as total labile P minus labile inorganic P.

SOIL ENZYME ASSAYS

Phosphomonoesterase, phosphodiesterase and pyrophosphatase enzyme assays were conducted according to procedures outlined by Tabatabai (1994). A procedure for determining phytase enzyme activity was adapted from Engelen et al. (2001) and outlined as follows. Each soil sample was analyzed in duplicate along with two blanks. One gram of soil was placed in a 50-mL Erlenmeyer flask with 4 mL of a pH 5.5 soil buffer and 0.2 mL of toluene. The flasks were then incubated for 5 minutes at 37° C. After 5 minutes, 10 mL of phytic acid substrate was added to each sample and mixed on a vortex mixer. Samples were then incubated again at 37° C for 65 minutes. Immediately after incubation 4 mL of color stop mixture were added and swirled to mix. Samples were then filtered through Whatman #2 filter paper and analyzed at 415 nm with a 0.04 mg L^{-1} detection limit. Standards were prepared containing 0, 5, 10, 15, 20, 25 and 30 ppm P. Four mL of standard solution was placed into a 50-mL Erlenmeyer flask and incubated for 5 minutes at 37° C. After 5 minutes 10 mL of deionized water was added and incubated continued at 37° C for 65 minutes. The standards were then treated and analyzed in the same manner as the unknown samples.

DATA ANALYSIS

The data were statistically analyzed using the Proc GLM model in SAS version 9.1 (SAS Institute, 2002). Differences within each P fraction were examined using ANOVA (analysis of variance) at a probability level (*P*) of 0.10, with mean separation determined using Fisher's Protected least significant difference (LSD) procedure. Each fraction was analyzed as a whole to determine overall treatment and time significance (i.e. single versus repeated application) and also to determine if there was a change from 2003 to 2004 within each fraction. "Time" refers to the 11 year difference between single applications and repeated applications and the differences between 2003 and 2004 will be referred to as the "sampling period". The time between "sampling periods" allowed us to examine the more refined P movement, where the long- term period showed more general transformations. In order to identify associations between the various P fractionations, Pearson correlation matrices were created and examined using Excel and analyzed statistically at P=0.10.

CHAPTER 3
RESULTS AND DISCUSSION

GENERAL SOIL ANALYSES

Soil pH, EC, inorganic C, total C, and total N results were variable (Tables 3.1 and 3.2; Figures 3.1 through 3.5). The WTR did not affect pH, but did affect EC (Figures 3.1 and 3.2). Repeated co-application led to greater ECs in 2003. The EC decreased significantly over the sampling period for repeated co-applications while single co-applications were stable over the same period. Most likely the added salts from WTR and biosolids have been leached in the years immediately after application. Single application levels were very close to background EC measurements. These findings were similar to those found by Ippolito and Barbarick (2004) where EC had decreased to a value similar to untreated control soils within 12 years following cessation of biosolids over-application to dryland winter wheat.

Inorganic C, total C, and total N were only significant for time. Inorganic C concentrations were less than 0.05% in every treatment (1991 and 2002) and so essentially 100% of the total C measured was organic (Figure 3.3). Organic C and Total N were very similar in patterns and concentrations, indicating that the source of both is essentially soil organic matter (Figures 3.4 and 3.5).

SOIL TEST P

Olsen, Mehlich-III, and Bray-I P tests were performed based on the Colorado P Risk Index, with results varying for the three tests (Tables 3.3 and 3.4; Figures 3.6 through 3.8). The Olsen test did not perform well because the study soils were slightly acidic; the Olsen extract dissolves P from calcium phosphates which are more prevalent in basic soils. Mehlich-III and Bray-I P test concentrations would be considered a high or very high risk according to the Colorado P Risk Index. Bray-I demonstrated the most significant influence from WTR when compared to control soils (Figure 3.8). The single and repeated biosolids control plots had average P concentrations of over 6,000 mg kg^{-1} and 16,000 mg kg^{-1}, respectively. Soils treated with biosolids and WTR had concentrations less than 200 mg kg^{-1}, a significant decrease. While all P concentrations were considered high according to the Colorado P Risk Index, there is no surface water in the area and an assessment would not be necessary in this situation. If water were in close proximity, a water extract to predict P availability in these soils would be most appropriate (Ippolito and Barbarick, 2006; Basta, Zupancic, and Dayton, 2000).

INORGANIC SOIL P FRACTIONATION

The overall inorganic P fraction decreased during this study period, with notable transformations with each fraction (Tables 3.5 and 3.6; Figures 3.9 through 3.13). The soluble and loosely bound inorganic fraction showed significant differences for treatment in the 2004 sampling year, with increasing WTR applications causing a reduction in this phase (Figure 3.9). The trend was the same in 2003, although not statistically significant. There were also noticeable, although not

Table 3.1
Average soil pH, EC, inorganic C, total C and total N from 0-5-cm depth measured in 2003 (n=4)

Treatment (mg ha^{-1})	Time (single/repeated)	pH	EC (dS m^{-1})	Inorganic C (g kg^{-1})	Total C %	Total N %
5	Single	5.4	0.37	0.000	2.35	0.21
10	Single	5.5	0.44	0.002	2.37	0.21
21	Single	5.5	0.42	0.000	2.19	0.19
Single application	Significance*	NS†	NS	NS	NS	NS
	LSD					
5	Repeated	5.6	1.31	0.003	3.06	0.29
10	Repeated	5.8	1.41	0.033	2.97	0.27
21	Repeated	5.9	1.55	0.045	3.13	0.29
Repeated application	Significance*	NS	NS	NS	NS	NS
	LSD					
Single vs repeated	Significance*	•	•	•	•	•
	LSD	0.29	0.31	0.31	0.13	0.02
Treatment by time interaction	Significance*	NS	NS	NS	NS	NS
	LSD					

* Significance (•) measured at the 10% probability level as determined by Fisher's Protected least significant difference (LSD) procedure.
† NS = non-significant.

always significant, decreases in the soluble/exchangeable phase over the sampling period. This change was likely due to increased plant growth requiring more P with the increase in precipitation from 2003 to 2004, and the soluble/exchangeable fraction was the most readily available. It may have been possible that P from the organic biomass fraction, an easily decomposable fraction, entered the soluble inorganic fraction and either went to plant production or was adsorbed by the non-crystalline (oxalate-extractable) fraction. In 2004 a significant treatment effect was observed, and 21 Mg ha^{-1} of WTR caused a P reduction in the single application to a concentration similar to that of untreated control soil.

The Al, Fe, and occluded inorganic phases were significant for time, indicating the repeated application led to increased P in these fractions (Figures 3.10 through 3.12). None of the fractions showed a significant treatment effect. The three fractions did not change significantly over the sampling period which indicates that, while there may be slow shifts, these three fractions are relatively stable. The observed increase in the Al and Fe fractions may be attributable to the repeated biosolids application. As found in a study by Maguire, Sims, and Coale (2000), biosolids additions led to increases in both Al and Fe-bound P when compared to untreated control soils. In comparison, occluded P levels can be attributed to WTR. While background occluded P concentrations were present, and some increases were associated with biosolids addition alone, there were overall increases observed with the addition of WTR.

Table 3.2
Average soil pH, EC, inorganic C, total C and total N from 0-5-cm depth measured in 2004 (n=4)

Treatment (mg ha^{-1})	Time (single/repeated)	pH	EC (dS m^{-1})	Inorganic C (g kg^{-1})	Total C %	Total N %
0	Untreated	5.68	0.36	0.00	1.41	0.12
10 biosolids only	Single	5.38	0.35	0.00	1.01	0.08
5	Single	5.20	0.41	0.02	1.96	0.19
10	Single	5.4	0.40	0.000	1.78	0.17
21	Single	5.3	0.33	0.000	1.48	0.14
Single application	Significance*	•	NS†	NS	NS	NS
	LSD	0.12				
Single co-application vs single biosolids	Significance*	•	NS	NS	NS	•
	LSD	0.14				0.08
0	Untreated	5.68	0.36	0.00	1.41	0.12
10 biosolids only	Repeated	5.45	0.54	0.00	1.94	0.18
5	Repeated	5.38	0.64	0.00	2.56	0.26
10	Repeated	5.3	0.81	0.017	2.38	0.23
21	Repeated	5.7	0.86	0.017	2.39	0.23
Repeated application	Significance*	NS	NS	NS	NS	NS
	LSD					
Repeated Co-application vs repeated biosolids	Significance*	NS	•	NS	NS	NS
	LSD		0.19			
Single vs repeated	Significance*	NS	•	NS	•	•
	LSD		0.31		0.68	0.06
Treatment by time Interaction	Significance*	NS	NS	NS	NS	NS
	LSD					
Single co-application vs. untreated	Significance*	•	NS	NS	NS	NS
	LSD	0.09				
Repeated co-application vs. untreated	Significance*	NS	•	NS	•	•
	LSD		0.27		0.67	0.07

* Significance (•) measured at the 10% probability level as determined by Fisher's Protected least significant difference (LSD) procedure.
† NS = non-significant.

Table 3.3
Average Olsen, Mehlich-III, and Bray-I P from the 0-5-cm depth measured in 2003 (n=4)

Treatment ($Mg\ ha^{-1}$)	Time (single/repeated)	Olsen P ($mg\ kg^{-1}$)	Mehlich-III P ($mg\ kg^{-1}$)	Bray-I P ($mg\ kg^{-1}$)
5	Single	160	171	121
10	Single	141	133	112
21	Single	139	131	104
Single application	Significance*	NS†	NS	NS
	LSD			
5	Repeated	209	226	144
10	Repeated	159	225	124
21	Repeated	133	202	123
Repeated application	Significance*	•	NS	NS
	LSD	41		
Single vs repeated	Significance*	NS	•	•
	LSD		21	7
Treatment by time interaction	Significance*	NS	NS	NS
	LSD			

* Significance (•) measured at the 10% probability level as determined by Fisher's Protected least significant difference (LSD) procedure.
† NS = non-significant.

The inorganic Ca-bound P fraction was not significant for WTR treatment or time (Figure 3.13). A decrease between sampling periods could be attributed to increased rainfall in 2004 which caused Ca-P soil phases to dissolve. The dissolution of Ca-P phases caused P to move into the Fe-bound and probably the Al-oxalate fractions. In support, changes in the Ca-bound fraction in the sampling period were associated with changes in the Fe-bound fraction based on Pearson correlations. Although we would not expect Ca-P phases to be present in the study soil due to pH < 7, Ca was inherent to the biosolids and thus was added to the soil with each biosolids addition. In addition, soil from the control plots contained a moderate level of Ca.

OXALATE EXTRACTABLE FRACTION AND PSI

The oxalate extractable fraction includes P adsorbed to amorphous, non-crystalline Al and Fe particles; the inorganic fraction previously described only measured associations between P and crystalline Al and Fe particles. The non-crystalline fraction is one of the most important inorganic fractions in this study because the majority of WTR-borne Al and Fe are in the amorphous, non-crystalline form.

Table 3.4
Average Olsen, Mehlich-III, and Bray-I P from the 0-5-cm depth measured in 2004 (n=4)

Treatment (Mg ha^{-1})	Time (single/repeated)	Olsen P (mg kg^{-1})	Mehlich-III P (mg kg^{-1})	Bray-I P (mg kg^{-1})
0	Untreated	18.7	0.00	0.00
10 biosolids only	Single	48.2	140	6300
5	Single	86.3	139	143
10	Single	70.6	113	123
21	Single	55.1	119	121
Single application	Significance*	•	NS†	NS
	LSD	20.9		
Single co-application vs single biosolids	Significance*	NS	NS	•
	LSD			2450
0	Untreated	18.7	0.00	0.00
10 biosolids only	Repeated	69.5	227	16400
5	Repeated	106	207	188
10	Repeated	61.5	186	160
21	Repeated	71.8	201	156
Repeated application	Significance*	NS	NS	NS
	LSD			
Repeated co-application vs repeated biosolids	Significance*	NS	NS	•
	LSD			5560
Single vs repeated	Significance*	NS	•	•
	LSD		30.4	26.6
Treatment by time interaction	Significance*	NS	NS	NS
	LSD			
Single co-application vs. untreated	Significance*	•	•	•
	LSD	17.7	30.7	34.2
Repeated co-application vs. untreated	Significance*	•	•	•
	LSD	29.6	35.4	23.7

* Significance (*) measured at the 10% probability level as determined by Fisher's Protected least significant difference (LSD) procedure.
† NS = non-significant.

Table 3.5
Average inorganic P fractions from the 0-5-cm depth measured in 2003 (n=4)

Treatment ($Mg\ ha^{-1}$)	Time (single/repeated)	Soluble and loosely bound ($mg\ kg^{-1}$)	Al-bound ($mg\ kg^{-1}$)	Fe-bound ($mg\ kg^{-1}$)	Occluded ($mg\ kg^{-1}$)	Ca-bound ($mg\ kg^{-1}$)
5	Single	31.1	16.8	103	74.9	186
10	Single	15.1	15.3	107	70.6	163
21	Single	9.80	20.4	118	81.9	168
Single application	Significance*	NS†	NS	NS	NS	NS
	LSD					
5	Repeated	23.3	62.0	209	109	193
10	Repeated	25.7	65.5	210	113	185
21	Repeated	15.7	68.9	197	150	157
Repeated application	Significance*	NS	NS	NS	•	NS
	LSD				32.0	
Single vs repeated	Significance*	NS	•	•	•	NS
	LSD		16.2	43.2	22.2	
Treatment by time interaction	Significance*	NS	NS	NS	NS	NS
	LSD					

* Significance (•) measured at the 10% probability level as determined by Fisher's Protected least significant difference (LSD) procedure.
† NS = non-significant.

The oxalate fraction showed significant increases from 2003 to 2004 as well as significant increases from single versus repeated applications. (Tables 3.7 and 3.8; Figures 3.14 through 3.16). Possibly the increased 2004 precipitation led to increased weathering and a corresponding increase in WTR surface area. A subsequent increase in WTR exposed surface area would lead to increased exchange sites for P sorption to occur.

The Al-oxalate fraction was statistically significant for treatment only in 2004 and showed a trend for time with no statistical significance. Both Fe and P-oxalate fractions were significant for time. All three fractions showed significant increases over the sampling period. Oxalate extractable Al increased as WTR application increased, indicating greater concentrations of amorphous Al additions. The WTR effect was prevalent when compared to the 2004 control plots which contained very low amorphous Al concentrations. Data from the 2004 control plots indicated amorphous Fe was contained naturally in the soil but was also added as a result of biosolids and WTR application. The city of Fort Collins, CO wastewater facility routinely adds $FeCl_3$ during the biosolids treatment process to control H_2S formation. This would contribute to the increase in amorphous Fe content in the co-applied plots. Measurements from the control plots in 2003 would have been helpful in this situation to observe changes in absence of WTR, and unfortunately were not obtained in this study.

Table 3.6
Average inorganic P fractions from the 0-5-cm depth measured in 2004 (n=4)

Treatment (Mg ha^{-1})	Time (single/repeated)	Soluble and loosely bound (mg kg^{-1})	Al-bound (mg kg^{-1})	Fe-bound (mg kg^{-1})	Occluded (mg kg^{-1})	Ca-bound (mg kg^{-1})
0	Untreated	1.26	0.00	43.7	17.8	112
10 biosolids only	Single	15.0	12.5	100	22.5	131
5	Single	10.9	50.3	129	59.8	141
10	Single	8.80	23.5	126	50.5	131
21	Single	1.90	12.7	117	53.6	136
Single application	Significance*	•	NS†	NS	•	NS
	LSD	4.94			5.65	
Single co-application vs single biosolids	Significance*	•	NS	NS	•	NS
	LSD	4.36			6.73	
0	Untreated	1.26	0.00	43.7	17.8	112
10 biosolids only	Repeated	22.5	57.9	245	35.7	163
5	Repeated	13.5	44.2	295	75.6	146
10	Repeated	7.80	58.8	239	74.4	146
21	Repeated	4.40	77.5	356	110	184
Repeated application	Significance*	•	NS	NS	NS	•
	LSD	5.47				29.4
Repeated co-application vs repeated biosolids	Significance*	•	NS	NS	•	•
	LSD	5.11			40.8	27.9
Single vs repeated	Significance*	NS	•	•	•	•
	LSD		30.2	79.9	20.6	13.5
Treatment by time interaction	Significance*	NS	NS	NS	NS	NS
	LSD					
Single co-application vs. untreated	Significance*	•	NS	•	•	•
	LSD	5.03		27.7	7.30	14.5
Repeated co-application vs. untreated	Significance*	•	•	•	•	•
	LSD	4.85	34.8	120	40.8	25.7

* Significance (•) measured at the 10% probability level as determined by Fisher's Protected least significant difference (LSD) procedure.
† NS = non-significant.

Table 3.7
Average oxalate-extractable P, Al and Fe, and PSI from the 0-5-cm depth measured in 2003 (n=4)

Treatment ($Mg\ ha^{-1}$)	Time (single/repeated)	P ($mg\ kg^{-1}$)	Al ($mg\ kg^{-1}$)	Fe ($mg\ kg^{-1}$)	PSI ($mmol\ kg^{-1}$)
5	Single	44.8	317	277	0.07
10	Single	46.4	661	257	0.05
21	Single	54.0	1180	274	0.05
Single application	Significance*	NS†	NS	NS	NS
	LSD				
5	Repeated	224	945	422	0.17
10	Repeated	277	2470	482	0.11
21	Repeated	205	2550	389	0.07
Repeated application	Significance*	•	•	•	•
	LSD	34.2	961	49.0	0.02
Single vs repeated	Significance*	NS	NS	NS	NS
	LSD				
Treatment by time interaction	Significance*	•	NS	•	•
	LSD				

* Significance (•) measured at the 10% probability level as determined by Fisher's Protected least significant difference (LSD) procedure.
† NS = non-significant.

The oxalate extraction also determines the amount of P associated with amorphous Fe and Al. The P-oxalate fraction was variable in 2003 and was significant for time in 2004. There were no observable trends indicating a treatment effect and most concentrations were similar to those soils which received only biosolids. Both biosolids control applications had significantly greater P concentrations as compared to soil that received no treatments, indicating that biosolids are the main contributor of P in the co-applied soils. This fraction also significantly increased over the sampling period suggesting a shift of P from the organic and other inorganic fractions to the oxalate-extractable fraction. These findings are consistent with other studies (Makris et al., 2004; Ippolito et al., 2003; Dayton and Basta 2005) that have observed significant P sorption to amorphous Al in WTR.

Because biosolids contain inherently large quantities of P, there is a risk of P saturation in treated soils. The PSI was calculated (equation 1) on all treatment soils and the three additional control soils (Tables 3.7 and 3.8; Figure 3.17). The 1991 biosolids application (with no WTR) had a PSI of 0.223; a repeated biosolids application increased the PSI to 0.312. Both treatments affected the PSI, were well below saturation, and the repeated biosolids application caused a PSI increase which would most likely continue given future applications.

Table 3.8
Average oxalate-extractable P, Al and Fe, and PSI from the 0-5-cm depth measured in 2004 (n=4)

Treatment (Mg ha^{-1})	Time (single/repeated)	P (mg kg^{-1})	Al (mg kg^{-1})	Fe (mg kg^{-1})	PSI (mg kg^{-1})
0	Untreated	97.5	768	676	0.08
10 biosolids only	Single	338	989	730	0.22
5	Single	343	1920	902	0.13
10	Single	319	2110	831	0.11
21	Single	327	3540	826	0.08
Single application	Significance*	NS†	•	NS	•
	LSD		929		0.03
Single co-application vs single biosolids	Significance*	NS	•	•	•
	LSD		778	65.2	0.04
0	Untreated	97.5	768	676	0.08
10 biosolids only	Repeated	755	1510	1140	0.31
5	Repeated	841	3430	1320	0.18
10	Repeated	713	4600	1220	0.14
21	Repeated	918	11100	1490	0.07
Repeated application	Significance*	NS	•	NS	•
	LSD		3210		0.04
Repeated co-application vs repeated biosolids	Significance*	NS	•	NS	•
	LSD		3000		0.04
Single vs repeated	Significance*	•	NS	•	NS
	LSD	194		189	
Treatment by time interaction	Significance*	NS	•	NS	NS
	LSD				
Single co-application vs. untreated	Significance*	•	•	•	•
	LSD	52.2	781	85.7	0.03
Repeated co-application vs. untreated	Significance*	•	•	•	•
	LSD	268	2920	247	0.04

* Significance (•) measured at the 10% probability level as determined by Fisher's Protected least significant difference (LSD) procedure.
† NS = non-significant.

The single and repeated additions of 21 Mg ha^{-1} of WTR reduced the PSI to 0.073 and 0.070, respectively, both lower than untreated control plot. The 10 Mg ha^{-1} and 5 Mg ha^{-1} of WTR also reduced PSI below the untreated control, but the decrease was not as dramatic as the 21 Mg ha^{-1} rate. Elliott et al. (2002) also observed decreases in PSI with WTR addition. Thus, the addition of WTR to these systems can prolong their life expectancy with regards to long-term biosolids application and P-loading.

ORGANIC SOIL P FRACTIONATION

The overall organic P fraction increased over the sampling period and showed significant differences in the single versus repeated applications (Tables 3.9 and 3.10; Figures 3.18 through 3.22). Changes over the sampling period may be attributed to either biosolids addition or increased precipitation during 2004, promoting increased plant growth and thus increasing the organic P fraction. Similarly, Sui, Thompson, and Shang (1999) was not able to absolutely distinguish between organic P increases from biosolids additions versus additions from increased plant production as a result of the biosolids.

Labile organic P was the only fraction that showed observable trends for treatment (Figure 3.18). Increasing WTR applications led to P decreases in the labile fraction. Phosphorus reductions were observed in WTR-treated when compared to control or biosolids only plots. The plot that received no treatments did not have any P in the labile or biomass fraction.

Biomass P was the only fraction that showed a decrease over the sampling period, while the other fractions increased (Figure 3.19). Biomass P is an easily labile organic fraction which can release P to the inorganic fraction. The decrease over the sampling period could be due to the loss of P to both the inorganic fraction as well as the amorphous, non-crystalline fraction.

Labile organic P showed a statistically significant increase in repeated applications versus single applications in 2003 and was variable in 2004 (Figure 3.18), most likely due to the recent addition of biosolids adding this fraction to the system. The moderately labile and humic acid fractions were both significant for time in 2003 and 2004 (Figures 3.20 and 3.22, respectively). The single and repeated co-application moderately labile fractions were also greater than the control but similar to biosolids addition alone, suggesting biosolids addition under co-application conditions contributed to the moderately labile organic P fraction. Fulvic acid was significant for time only in 2003 and was reduced to below detection limits in 2004 (Figure 3.21). This decrease may be associated with the increased 2004 precipitation, increased weathering, and thus increased microbial activity.

P CYCLING

Using a Pearson correlation matrix (Table 3.11) in conjunction with ANOVA, we established relationships between changes in each of the P fractions. From these relationships Figure 3.23 was formed to illustrate the shifts in P forms within this study.

Data revealed increases and decreases in specific fractions, and correlation matrices allowed us to predict fractions that moved simultaneously indicating P cycling pathways. Microbial biomass P was the only organic fraction that decreased from 2003 to 2004, with the decrease correlated with a decrease in labile organic P and indirectly correlated with an increase in soluble/exchangeable inorganic P in the same time period. This suggests a path of P cycling between these fractions.

Table 3.9
Average organic P fractions from the 0-5-cm depth measured in 2003 (n=4)

Treatment ($Mg\ ha^{-1}$)	Time (single/repeated)	Labile Po ($mg\ kg^{-1}$)	Biomass Po ($mg\ kg^{-1}$)	Mod. Labile Po ($mg\ kg^{-1}$)
5	Single	165	623	585
10	Single	138	393	549
21	Single	181	397	737
Single application	Significance*	NS†	NS	NS
	LSD			
5	Repeated	353	1100	2000
10	Repeated	434	697	2150
21	Repeated	332	569	2320
Repeated application	Significance*	NS	•	NS
	LSD		304	
Single vs repeated	Significance*	•	•	•
	LSD	206	248	759
Treatment by time interaction	Significance*	NS	NS	NS
	LSD			
		Fulvic acid Po ($mg\ kg^{-1}$)	**Humic acid Po ($mg\ kg^{-1}$)**	**Non-labile Po ($mg\ kg^{-1}$)**
5	Single	399	292	5.35
10	Single	307	348	5.00
21	Single	208	443	10.5
Single application	Significance*	•	NS	NS
	LSD	109		
5	Repeated	426	506	16.3
10	Repeated	376	491	2.63
21	Repeated	555	396	6.85
Repeated application	Significance*	NS	NS	•
	LSD			8.32
Single vs repeated	Significance*	•	•	NS
	LSD	137	81.7	
Treatment by time interaction	Significance*	NS	NS	NS
	LSD			

* Significance (•) measured at the 10% probability level as determined by Fisher's Protected least significant difference (LSD) procedure.
† NS = non-significant.

Table 3.10
Average organic P fractions from the 0-5-cm depth measured in 2004 (n=4)

Treatment ($Mg\ ha^{-1}$)	Time (single/repeated)	Labile Po ($mg\ kg^{-1}$)	Biomass Po ($mg\ kg^{-1}$)	Mod. Labile Po ($mg\ kg^{-1}$)
0	Untreated	20.3	0.00	370
10 biosolids only	Single	553	0.00	1040
5	Single	377	258	957
10	Single	451	198	991
21	Single	167	251	1280
Single application	Significance*	•	NS†	•
	LSD	165		232
Single co-application vs single biosolids	Significance*	•	NS	NS
	LSD	229		
0	Untreated	20.3	0.00	370
10 biosolids only	Repeated	1300	137	2060
5	Repeated	1232	190	2540
10	Repeated	757	101	2140
21	Repeated	634	382	2490
Repeated application	Significance*	•	NS	NS
	LSD	338		
Repeated co-application vs repeated biosolids	Significance*	NS	NS	NS
	LSD			
Single vs repeated	Significance*	•	NS	•
	LSD	303		444
Treatment by time interaction	Significance*	NS	NS	NS
	LSD			
Single co-application vs. untreated	Significance*	•	NS	•
	LSD	170		260
Repeated co-application vs. untreated	Significance*	•	NS	•
	LSD	341		567

(continued)

Table 3.10 (Continued)

Treatment (Mg ha^{-1})	Time (single/repeated)	Fulvic Acid Po (mg kg^{-1})	Humic Acid Po (mg kg^{-1})	Non-labile Po (mg kg^{-1})
0	Untreated	0.00	392	5.67
10 biosolids only	Single	0.00	746	12.8
5	Single	0.00	482	7.02
10	Single	0.00	526	9.79
21	Single	0.00	632	8.28
Single application	Significance*	NS	NS	NS
	LSD			
Single co-application vs single biosolids	Significance*	NS	NS	NS
	LSD			
0	Untreated	0.00	392	5.67
10 biosolids only	Repeated	0.00	945	15.4
5	Repeated	0.00	1080	11.2
10	Repeated	0.00	755	9.37
21	Repeated	0.00	1066	16.2
Repeated application	Significance*	NS	•	NS
	LSD		230	
Repeated co-application vs repeated biosolids	Significance*	NS	•	NS
	LSD		200	
Single vs repeated	Significance*	NS	NS	NS
	LSD			
Treatment by time interaction	Significance*	NS	•	NS
	LSD			
Single co-application vs. untreated	Significance*	NS	NS	NS
	LSD			
Repeated co-application vs. untreated	Significance*	NS	•	NS
	LSD		323	

* Significance (•) measured at the 10% probability level as determined by Fisher's Protected least significant difference (LSD) procedure.
† NS = non-significant.

Table 3.11
Pearson correlation matrix for 2003 and 2004 data*

	Trt	Old/New	Soluble/ Exchangeable	Al-bound	Fe-bound	Occluded	Ca-bound	Olsen	Mehlich	Bray	P oxalate	Al oxalate	Fe oxalate	Labile Po	Biomass P
Trt	1.00														
Old/New	0.00	1.00													
Soluble/ exchangeable	-0.68	0.12	1.00												
Al-bound	-0.01	**0.45**	-0.03	1.00											
Fe-bound	0.12	**0.76**	0.04	**0.63**	1.00										
Occluded	0.23	**0.55**	-0.07	**0.66**	**0.92**	1.00									
Ca-bound	0.36	0.51	-0.10	0.54	0.78	0.84	1.00								
Olsen	**-0.45**	0.17	0.61	*0.40*	**0.36**	0.24	0.16	1.00							
Mehlich	-0.08	0.80	0.39	**0.50**	**0.80**	**0.69**	0.62	**0.38**	1.00						
Bray	-0.29	0.58	0.52	**0.37**	**0.58**	0.50	0.40	**0.46**	**0.89**	1.00					
P oxalate	0.55	0.52	-0.32	**0.49**	**0.72**	**0.70**	0.76	0.02	**0.44**	*0.41*	1.00				
Al oxalate	*0.40*	0.82	0.10	**0.61**	**0.96**	**0.87**	0.78	0.33	0.81	0.56	**0.76**	1.00			
Fe oxalate	0.07	0.83	0.15	**0.61**	**0.95**	**0.86**	0.76	0.30	**0.87**	0.65	**0.69**	**0.98**	1.00		
Labile Po	-0.38	0.64	0.43	0.40	**0.45**	0.26	0.25	0.24	**0.69**	**0.75**	*0.53*	0.51	0.59	1.00	
Biomass P	*-0.38*	*0.43*	-0.03	**0.45**	**0.40**	**0.46**	0.27	**0.44**	*0.06*	*0.55*	*0.50*	0.30	*0.49*	-0.47	1.00
Mod. labile Po	0.10	**0.88**	0.09	**0.54**	**0.88**	**0.72**	0.58	0.26	**0.85**	**0.62**	**0.62**	**0.88**	**0.92**	*0.48*	*0.44*
Humic acid	0.14	0.69	0.10	0.41	**0.65**	0.58	0.59	0.17	0.67	0.61	0.45	0.71	0.74	0.68	-0.01
Non-labile Po	0.18	0.25	-0.27	-0.07	0.29	0.22	0.33	*0.40*	0.12	0.13	0.44	0.32	0.23	*0.43*	-0.26
Monoesterase	-0.07	*-0.40*	0.08	-0.43	*-0.46*	*-0.37*	-0.19	-0.03	-0.39	-0.39	-0.32	-0.27	-0.27	-0.20	-0.03
Diesterase	0.05	0.13	0.04	-0.09	0.06	*0.37*	0.12	0.29	0.03	0.03	-0.06	0.07	0.10	0.09	*0.50*
Pyro	**-0.53**	-0.09	0.55	0.01	0.02	-0.22	*0.61*	0.23	0.13	0.17	-0.39	0.01	0.06	0.14	0.12
Phytase	-0.25	0.46	0.41	0.23	0.41	*0.44*	*-0.41*	0.20	0.68	0.81	-0.05	0.36	0.46	0.60	-0.06

* Italicized and underlined values denote significant 2003 or 2004 correlations, respectively. Bold values denote significant correlation in both years. Numerical data listed is for 2004 except for italicized cells which correspond to 2003 numerical data

The inorganic Al-bound, Fe-bound and occluded fractions showed little movement from 2003 and 2004, indicating stability within the fractions. Correlations indicated the Ca-bound fraction contributed P to the Fe-bound fraction and may contribute to P increases in Al-oxalate fraction. The soluble and loosely bound P fraction is also related to an increase in P in the Al-oxalate fraction, indicating that it may also be contributing to increases in P sorption by amorphous Al.

SOIL ENZYME ASSAYS

A major study objective was to determine if P mineralization rates could be calculated via enzyme assays. In order for mineralization rates to be measured there would need to be a shift from organic P to inorganic P; the opposite was observed in this study. Mineralization/ immobilization cycling constantly occurs in soil, and in this study there was an unquantifiable net immobilization effect. Only the organic biomass P fraction decreased, possibly shifting P to the inorganic fraction. The changes in the biomass P fraction over the sampling period were only correlated to the phosphodiesterase (diesterase) enzyme in one year.

Phosphomonoesterase (monoesterase) and diesterase were not statistically significant for time or treatment in 2003 or 2004 (Tables 3.12 and 3.13; Figures 3.24 and 3.25). Neither of the enzymes provided consistent observable trends and neither were consistently correlated to any changes in the organic P fractions. Monoesterase activity was significantly greater in the co-applied plots when compared to the control and biosolids alone plots. In fact, soils that did not receive WTR had little to no measured monoesterase activity. Monoesterase is known to be secreted by plants under soil P deficiency conditions, and therefore it is possible that any WTR additions may trigger a deficiency response, causing plants to secrete the monoesterase enzyme. The WTR caused an opposite reaction with diesterase, with the co-applied soils showing significant reductions in enzyme activity compared to the control soils.

The pyrophosphatase enzyme is one of the least studied enzymes of the phosphatase group. In this study the pyrophosphatase enzyme was not significant in 2003 and significant for treatment in 2004. Increased WTR application rates led to a decrease in pyrophosphatase activity (Tables 3.12 and 3.13; Figure 3.26). Changes from 2003 to 2004 were insignificant with the exception of one instance, but showed a decreasing activity trend. Single co-application treatments resulted in higher enzyme activity levels when compared to repeated co-applications. This could be due to a lack-of-need to degrade polyphosphates in the first few years following biosolids application because biosolids supplied necessary P.

The phytase enzyme also has not been extensively studied in soils but showed some promise for predicting several organic P fractions. Phytase activity was not significant in 2003 but was significant for time in 2004 (Tables 3.12 and 3.13; Figure 3.27). Correlations were variable but phytase activity was correlated to labile and moderately labile P in 2004 and correlated to humic acid P in both years. It is possible that the humic acid fraction has a consistent amount of substrate over time.

Table 3.12
Average P from monoesterase, diesterase, pyrophosphatase and phytase enzymes from the 0-5-cm depth measured in 2003 (n=4)

Treatment ($Mg\ ha^{-1}$)	Time (single/repeated)	Monoesterase P ($mg\ kg^{-1}$)	Diesterase P ($mg\ kg^{-1}$)	Pyrophosphatase P ($mg\ kg^{-1}$)	Phytase P ($mg\ kg^{-1}$)
5	Single	38.3	27.2	88.5	11.7
10	Single	33.9	23.0	37.2	14.6
21	Single	41.4	22.1	39.7	11.4
Single application	Significance*	NS†	•	NS	NS
	LSD		3.93		
5	Repeated	30.7	29.6	63.8	12.8
10	Repeated	33.9	25.8	50.0	13.5
21	Repeated	28.7	27.4	43.2	15.6
Repeated application	Significance*	NS	NS	NS	NS
	LSD				
Single vs repeated	Significance*	NS	NS	NS	NS
	LSD				
Treatment by time interaction	Significance*	NS	NS	NS	NS
	LSD				

* Significance (•) measured at the 10% probability level as determined by Fisher's Protected least significant difference (LSD) procedure.
† NS = non-significant.

Table 3.13
Average P from monoesterase, diesterase, pyrophosphatase and phytase enzymes from the 0-5-cm depth measured in 2004 (n=4)

Treatment (Mg ha^{-1})	Time (single/repeated)	Monoesterase P (mg kg^{-1})	Diesterase P (mg kg^{-1})	Pyrophosphatase P (mg kg^{-1})	Phytase P (mg kg^{-1})
0	Untreated	0.06	77.1	29.6	3.92
10 biosolids only	Single	0.07	101	47.5	2.53
5	Single	30.8	35.3	44.3	8.16
10	Single	31.1	42.4	34.2	8.38
21	Single	36.7	52.6	20.7	6.89
Single application	Significance*	•	NS†	NS	NS
	LSD	2.25			
Single co-application vs single biosolids	Significance*	•	NS	NS	•
	LSD	4.97			4.03
0	Untreated	0.06	77.1	29.6	3.92
10 biosolids only	Repeated	0	113	30.2	4.16
5	Repeated	33.7	59.1	51.5	12.3
10	Repeated	27.2	39.9	23.8	11.0
21	Repeated	24.3	42.9	15.5	9.50
Repeated application	Significance*	•	NS	•	NS
	LSD	7.00		26.2	
Repeated co-application vs repeated biosolids	Significance*	•	NS	•	•
	LSD	5.68		20.3	3.13
Single vs repeated	Significance*	NS	NS	NS	•
	LSD				3.04
Treatment by time interaction	Significance*	•	NS	NS	NS
	LSD				
Single co-application vs. untreated	Significance*	•	•	NS	NS
	LSD	5.07	24.7		
Repeated co-application vs. untreated	Significance*	•	•	NS	•
	LSD	5.68	23.0		25.0

* Significance (•) measured at the 10% probability level as determined by Fisher's Protected least significant difference (LSD) procedure.
† NS = non-significant.

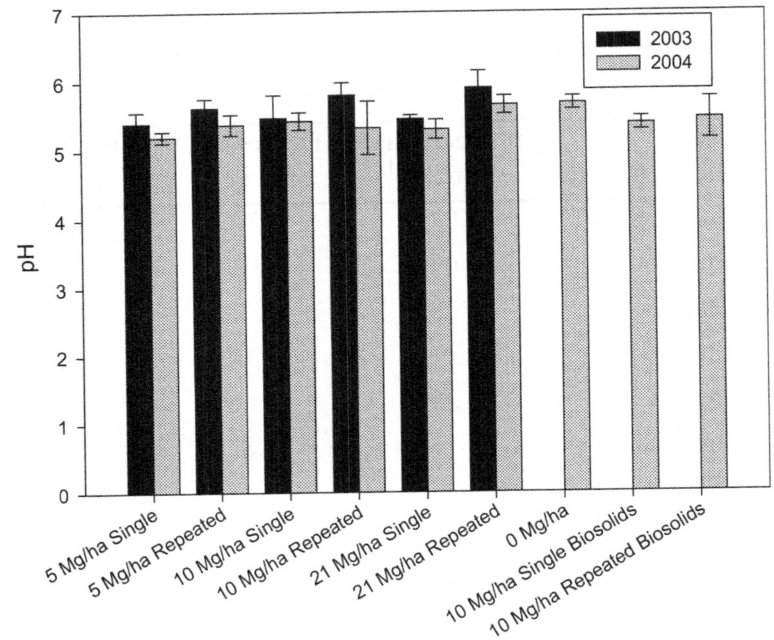

Figure 3.1 Soil pH from plots of 10 Mg ha^{-1} biosolids co-applied with 5, 10, or 21 Mg ha^{-1} WTR, 0-5-cm depth, 2003 and 2004. Single and repeated co-applications occurred in 1991 and 2002, respectively. Control soils, measured only in 2004, included an untreated control and a single and repeated biosolids application. Error bars represent one standard error of the mean (n=4).

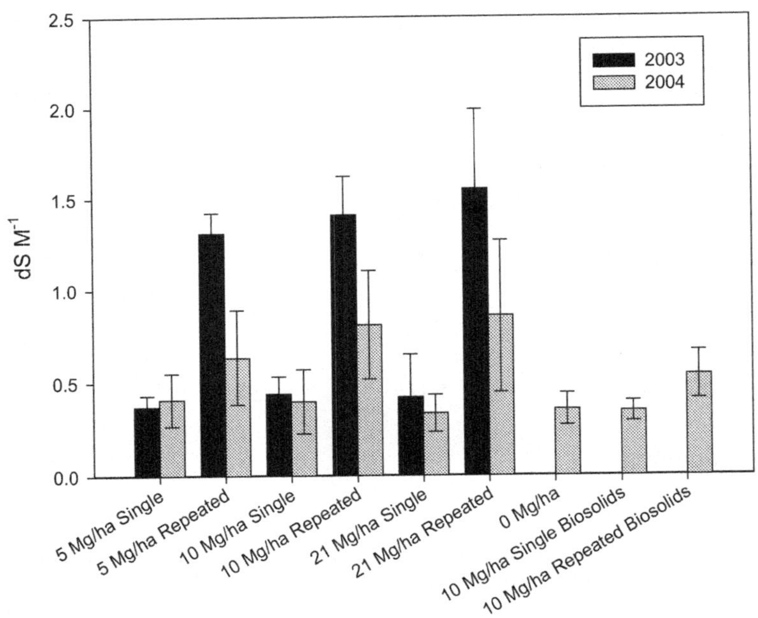

Figure 3.2 Soil EC from plots of 10 Mg ha^{-1} biosolids co-applied with 5, 10, or 21 Mg ha^{-1} WTR, 0-5-cm depth, 2003 and 2004. Single and repeated co-applications occurred in 1991 and 2002, respectively. Control soils, measured only in 2004, included an untreated control and a single and repeated biosolids application. Error bars represent one standard error of the mean (n=4).

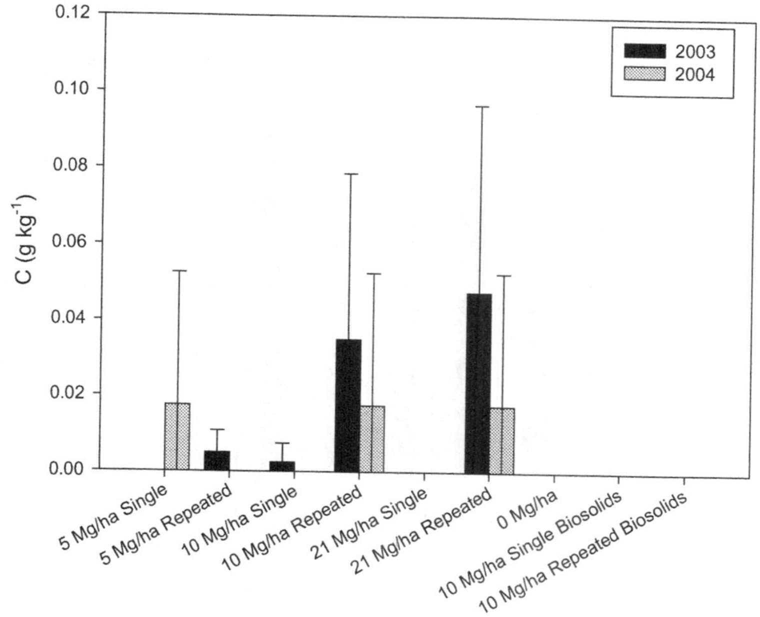

Figure 3.3 Inorganic soil C from plots of 10 Mg ha^{-1} biosolids co-applied with 5, 10, or 21 Mg ha^{-1} WTR, 0-5-cm depth, 2003 and 2004. Single and repeated co- applications occurred in 1991 and 2002, respectively. Control soils, measured only in 2004, included an untreated control and a single and repeated biosolids application. Error bars represent one standard error of the mean (n=4).

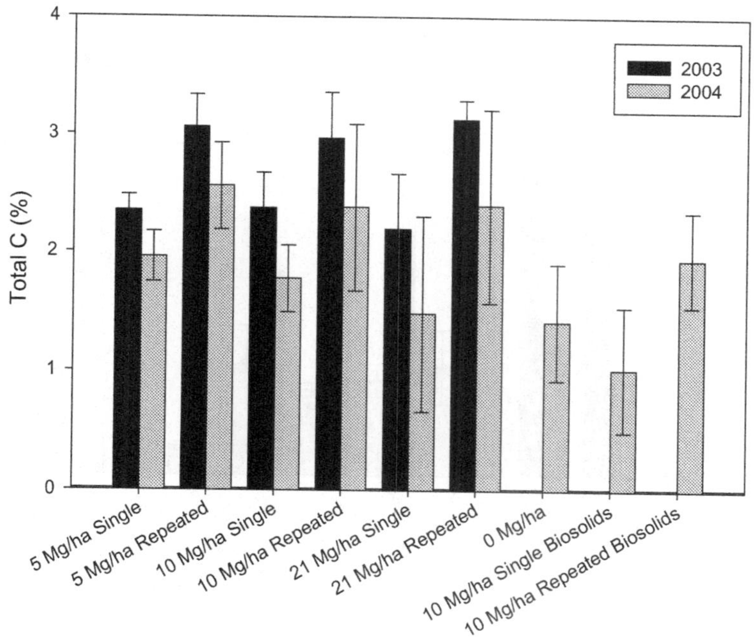

Figure 3.4 Total soil C from plots of 10 Mg ha^{-1} biosolids co-applied with 5, 10, or 21 Mg ha^{-1} WTR, 0-5-cm depth, 2003 and 2004. Single and repeated co-applications occurred in 1991 and 2002, respectively. Control soils, measured only in 2004, included an untreated control and a single and repeated biosolids application. Error bars represent one standard error of the mean (n=4).

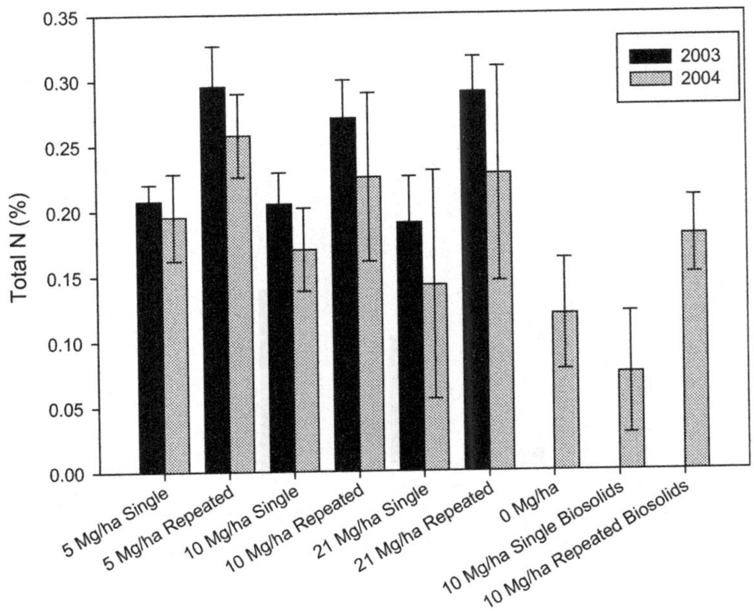

Figure 3.5 Total soil N from plots of 10 Mg ha^{-1} biosolids co-applied with 5, 10, or 21 Mg ha^{-1} WTR, 0-5-cm depth, 2003 and 2004. Single and repeated co-applications occurred in 1991 and 2002, respectively. Control soils, measured only in 2004, included an untreated control and a single and repeated biosolids application. Error bars represent one standard error of the mean (n=4).

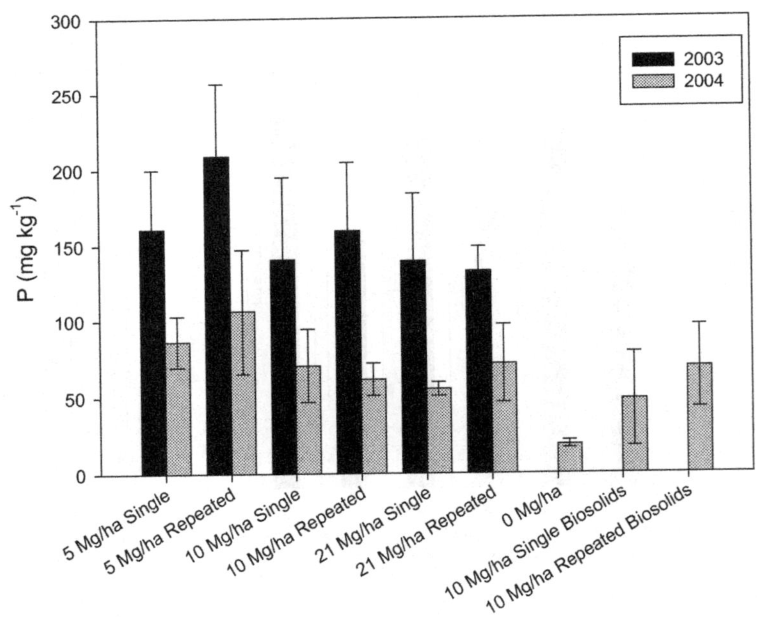

Figure 3.6 Olsen-extractable soil P from plots of 10 Mg ha^{-1} biosolids co-applied with 5, 10, or 21 Mg ha^{-1} WTR, 0-5-cm depth, 2003 and 2004. Single and repeated co-applications occurred in 1991 and 2002, respectively. Control soils, measured only in 2004, included an untreated control and a single and repeated biosolids application. Error bars represent one standard error of the mean (n=4).

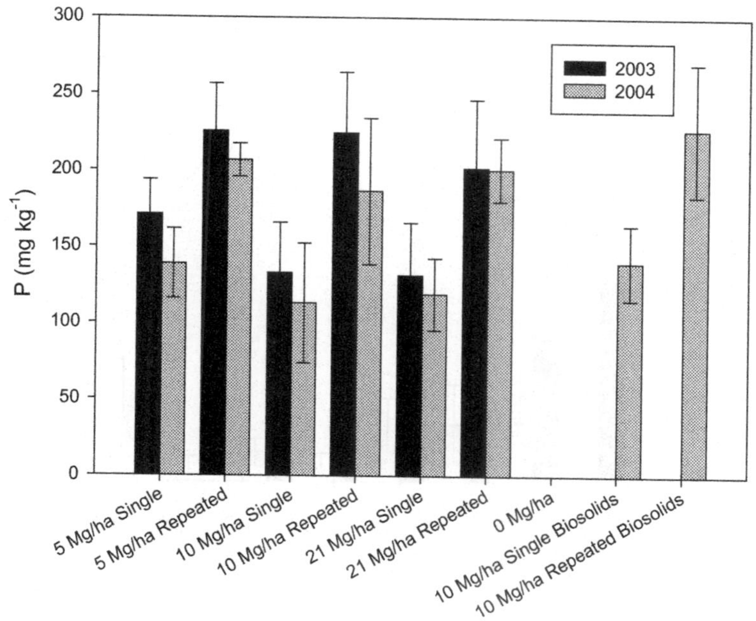

Figure 3.7 Mehlich-III-extractable soil P from plots of 10 Mg ha^{-1} biosolids co-applied with 5, 10, or 21 Mg ha^{-1} WTR, 0-5-cm depth, 2003 and 2004. Single and repeated co-applications occurred in 1991 and 2002, respectively. Control soils, measured only in 2004, included an untreated control and a single and repeated biosolids application. Error bars represent one standard error of the mean (n=4).

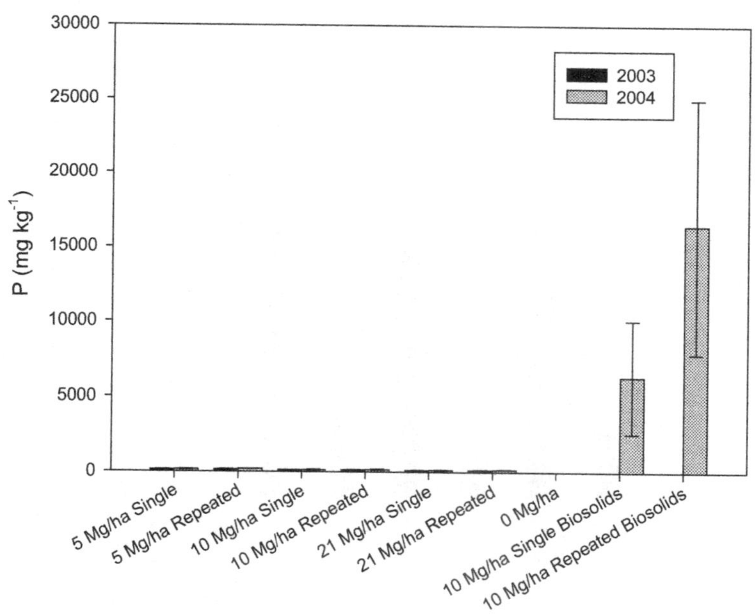

Figure 3.8 Bray-I-extractable soil P from plots of 10 Mg ha^{-1} biosolids co-applied with 5, 10, or 21 Mg ha^{-1} WTR, 0-5-cm depth, 2003 and 2004. Single and repeated co-applications occurred in 1991 and 2002, respectively. Control soils, measured only in 2004, included an untreated control and a single and repeated biosolids application. Error bars represent one standard error of the mean (n=4).

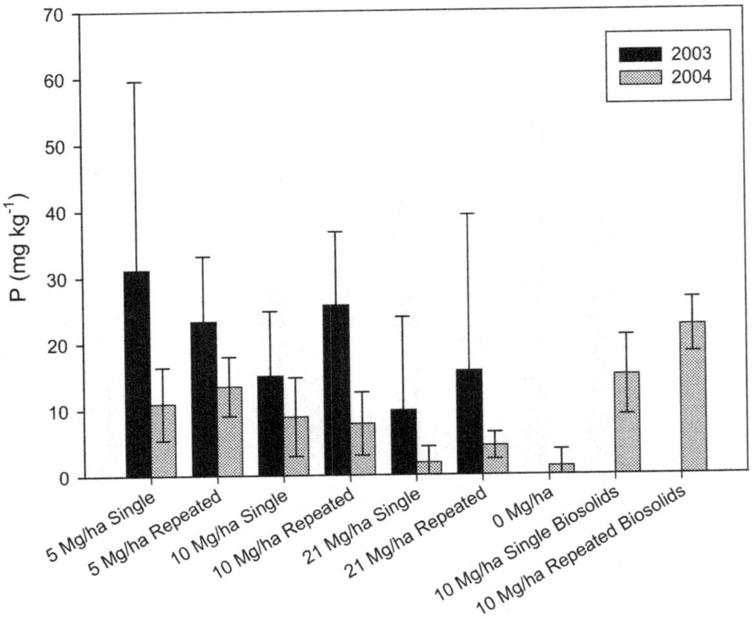

Figure 3.9 Soluble and loosely bound P from plots of 10 Mg ha^{-1} biosolids co-applied with 5, 10, or 21 Mg ha^{-1} WTR, 0-5-cm depth, 2003 and 2004. Single and repeated co-applications occurred in 1991 and 2002, respectively. Control soils, measured only in 2004, included an untreated control and a single and repeated biosolids application. Error bars represent one standard error of the mean (n=4).

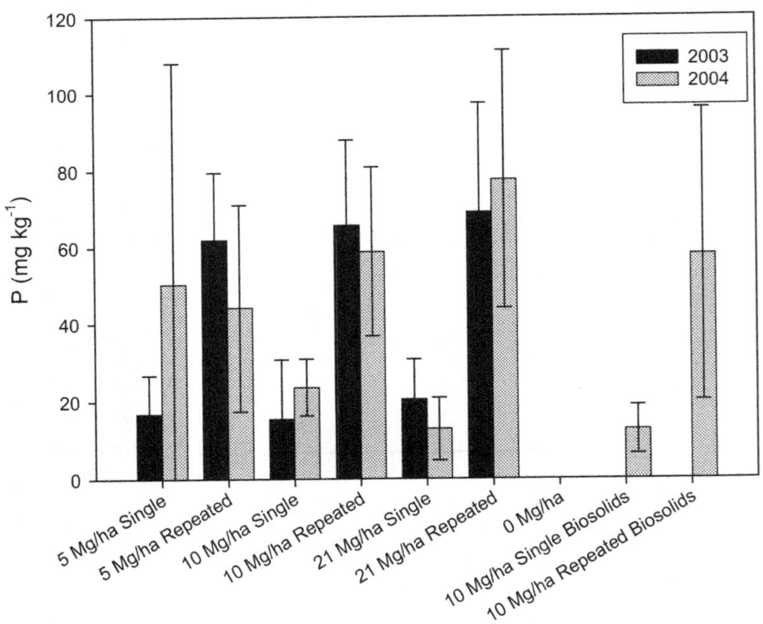

Figure 3.10 Aluminum bound P from plots of 10 Mg ha^{-1} biosolids co-applied with 5, 10, or 21 Mg ha^{-1} WTR, 0-5-cm depth, 2003 and 2004. Single and repeated co-applications occurred in 1991 and 2002, respectively. Control soils, measured only in 2004, included an untreated control and a single and repeated biosolids application. Error bars represent one standard error of the mean (n=4).

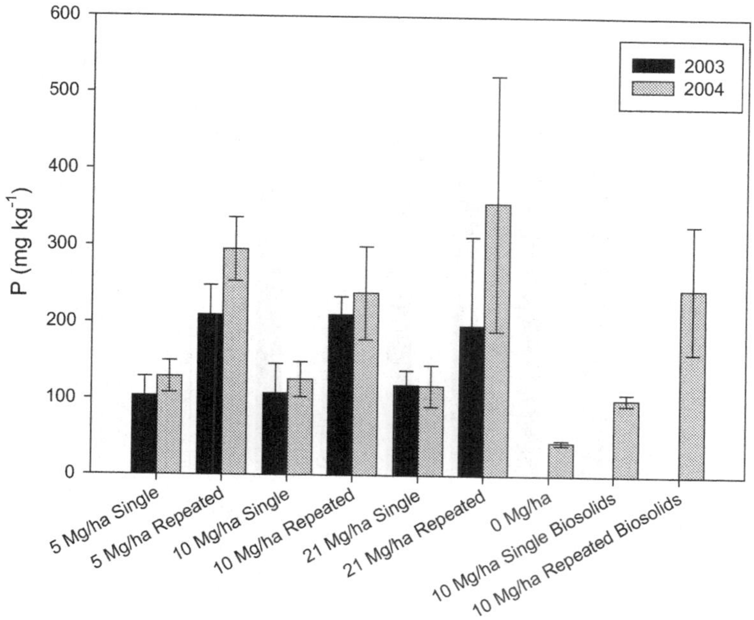

Figure 3.11 Iron bound P from plots of 10 Mg ha^{-1} biosolids co-applied with 5, 10, or 21 Mg ha^{-1} WTR, 0-5-cm depth, 2003 and 2004. Single and repeated co- applications occurred in 1991 and 2002, respectively. Control soils, measured only in 2004, included an untreated control and a single and repeated biosolids application. Error bars represent one standard error of the mean (n=4).

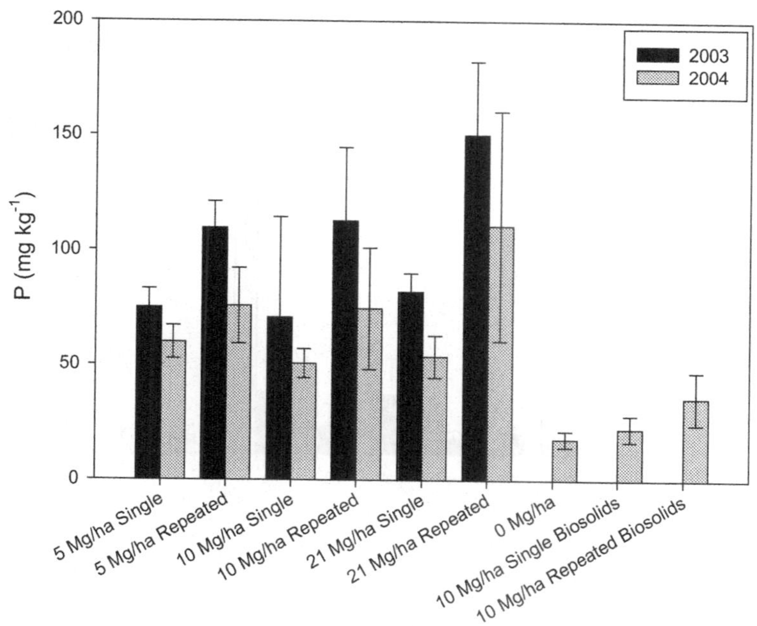

Figure 3.12 Occluded P from plots of 10 Mg ha^{-1} biosolids co-applied with 5, 10, or 21 Mg ha^{-1} WTR, 0-5-cm depth, 2003 and 2004. Single and repeated co-applications occurred in 1991 and 2002, respectively. Control soils, measured only in 2004, included an untreated control and a single and repeated biosolids application. Error bars represent one standard error of the mean (n=4).

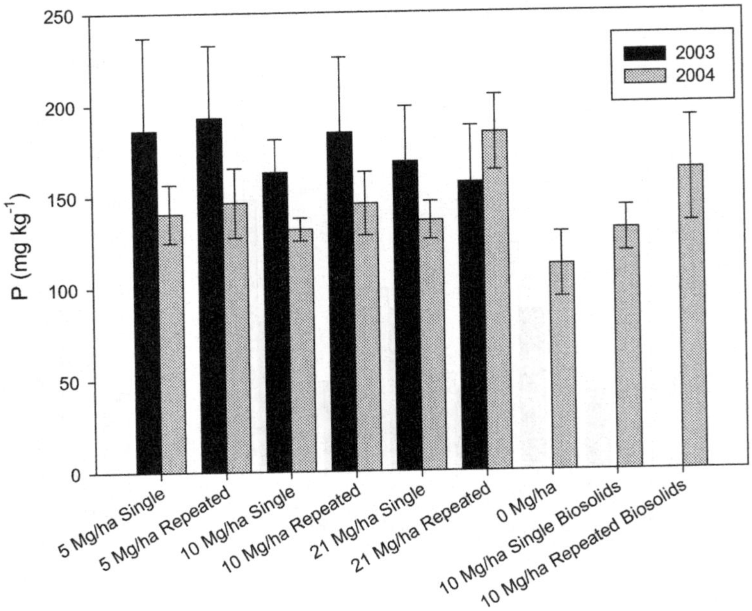

Figure 3.13 Calcium bound P from plots of 10 Mg ha^{-1} biosolids co-applied with 5, 10, or 21 Mg ha^{-1} WTR, 0-5-cm depth, 2003 and 2004. Single and repeated co- applications occurred in 1991 and 2002, respectively. Control soils, measured only in 2004, included an untreated control and a single and repeated biosolids application. Error bars represent one standard error of the mean (n=4).

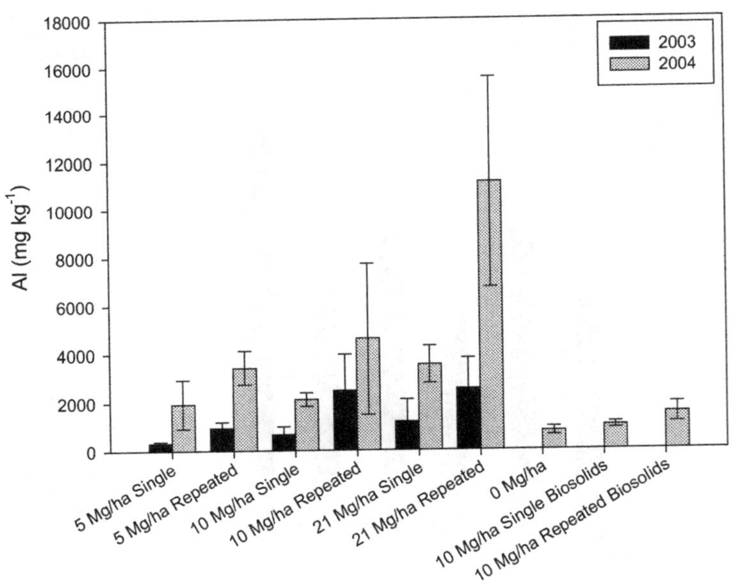

Figure 3.14 Oxalate-extractable Al from plots of 10 Mg ha^{-1} biosolids co-applied with 5, 10, or 21 Mg ha^{-1} WTR, 0-5-cm depth, 2003 and 2004. Single and repeated co-applications occurred in 1991 and 2002, respectively. Control soils, measured only in 2004, included an untreated control and a single and repeated biosolids application. Error bars represent one standard error of the mean (n=4).

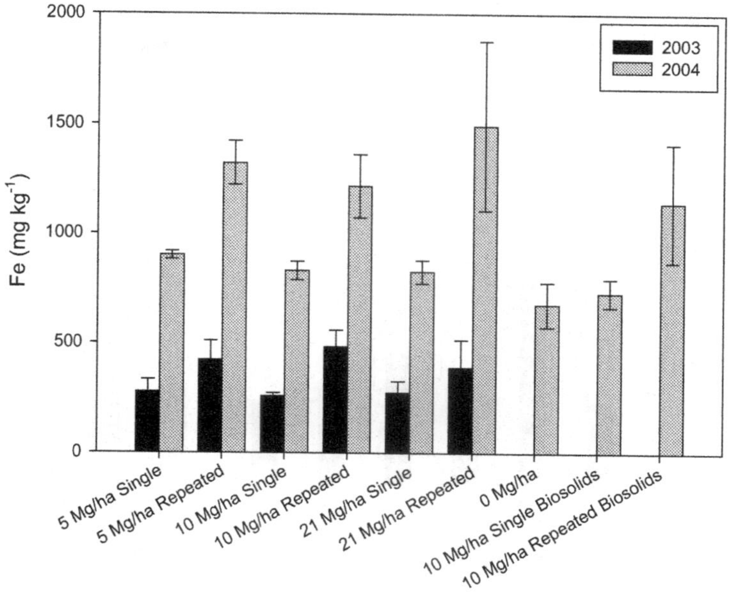

Figure 3.15 Oxalate-extractable Fe from plots of 10 Mg ha^{-1} biosolids co-applied with 5, 10, or 21 Mg ha^{-1} WTR, 0-5-cm depth, 2003 and 2004. Single and repeated co-applications occurred in 1991 and 2002, respectively. Control soils, measured only in 2004, included an untreated control and a single and repeated biosolids application. Error bars represent one standard error of the mean (n=4).

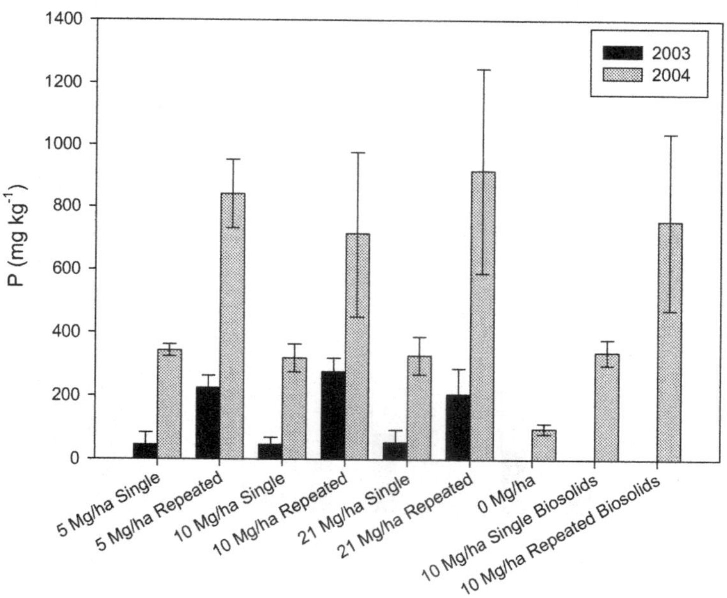

Figure 3.16 Oxalate-extractable P from plots of 10 Mg ha^{-1} biosolids co-applied with 5, 10, or 21 Mg ha^{-1} WTR, 0-5-cm depth, 2003 and 2004. Single and repeated co-applications occurred in 1991 and 2002, respectively. Control soils, measured only in 2004, included an untreated control and a single and repeated biosolids application. Error bars represent one standard error of the mean (n=4).

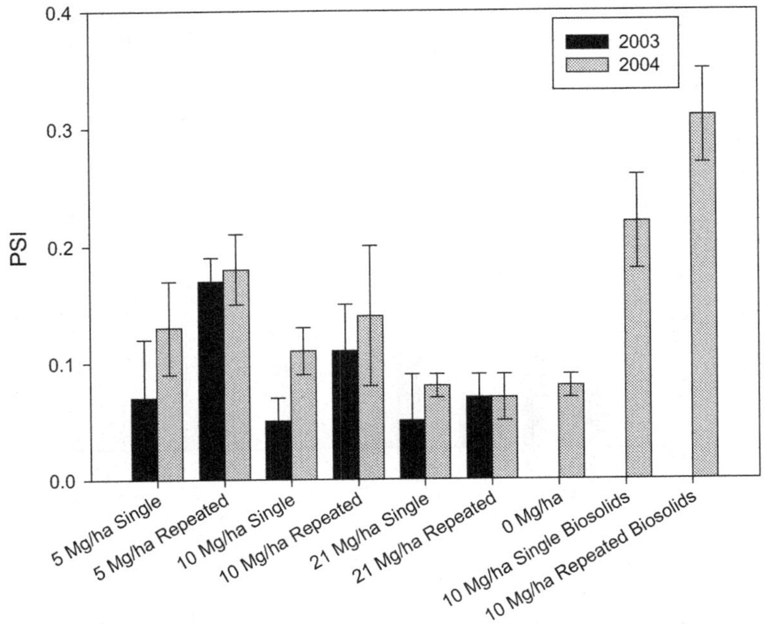

Figure 3.17 Phosphorus saturation index (PSI) from plots of 10 Mg ha^{-1} biosolids co-applied with 5, 10, or 21 Mg ha^{-1} WTR, 0-5-cm depth, 2003 and 2004. Single and repeated co-applications occurred in 1991 and 2002, respectively. Control soils, measured only in 2004, included an untreated control and a single and repeated biosolids application. Error bars represent one standard error of the mean (n=4).

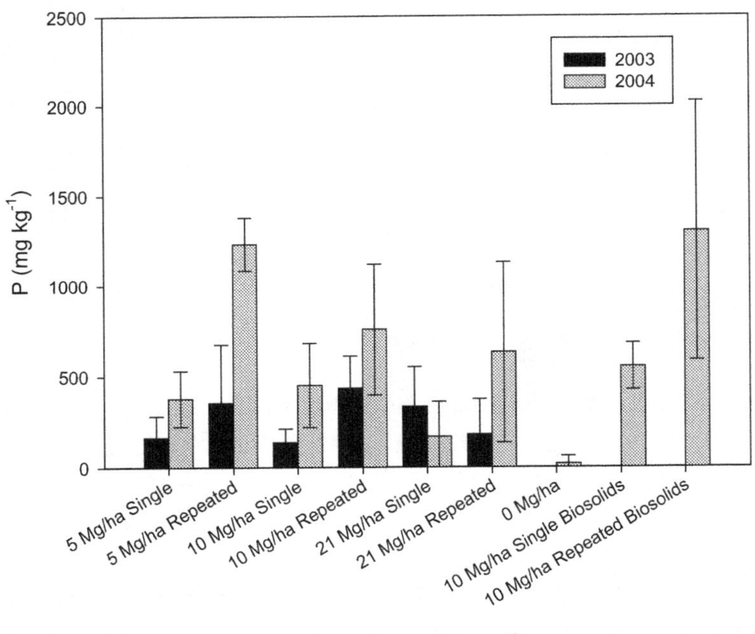

Figure 3.18 Labile organic P from plots of 10 Mg ha^{-1} biosolids co-applied with 5, 10, or 21 Mg ha^{-1} WTR, 0-5-cm depth, 2003 and 2004. Single and repeated co-applications occurred in 1991 and 2002, respectively. Control soils, measured only in 2004, included an untreated control and a single and repeated biosolids application. Error bars represent one standard error of the mean (n=4).

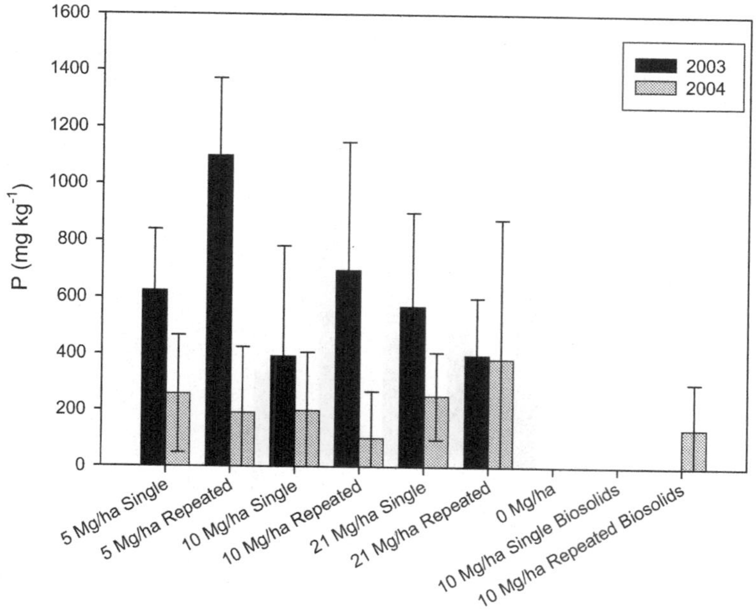

Figure 3.19 Soil microbial biomass P from plots of 10 Mg ha^{-1} biosolids co-applied with 5, 10, or 21 Mg ha^{-1} WTR, 0-5-cm depth, 2003 and 2004. Single and repeated co-applications occurred in 1991 and 2002, respectively. Control soils, measured only in 2004, included an untreated control and a single and repeated biosolids application. Error bars represent one standard error of the mean (n=4).

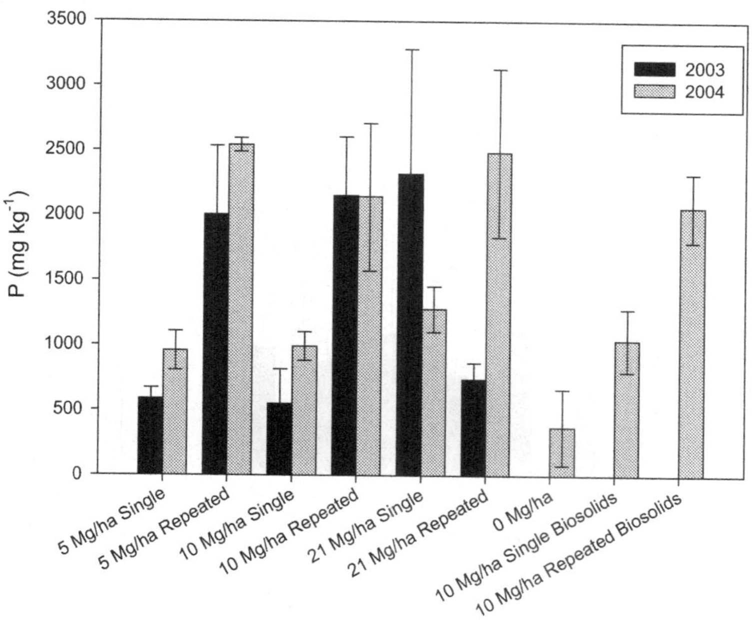

Figure 3.20 Moderately labile organic P from plots of 10 Mg ha^{-1} biosolids co-applied with 5, 10, or 21 Mg ha^{-1} WTR, 0-5-cm depth, 2003 and 2004. Single and repeated co-applications occurred in 1991 and 2002, respectively. Control soils, measured only in 2004, included an untreated control and a single and repeated biosolids application. Error bars represent one standard error of the mean (n=4).

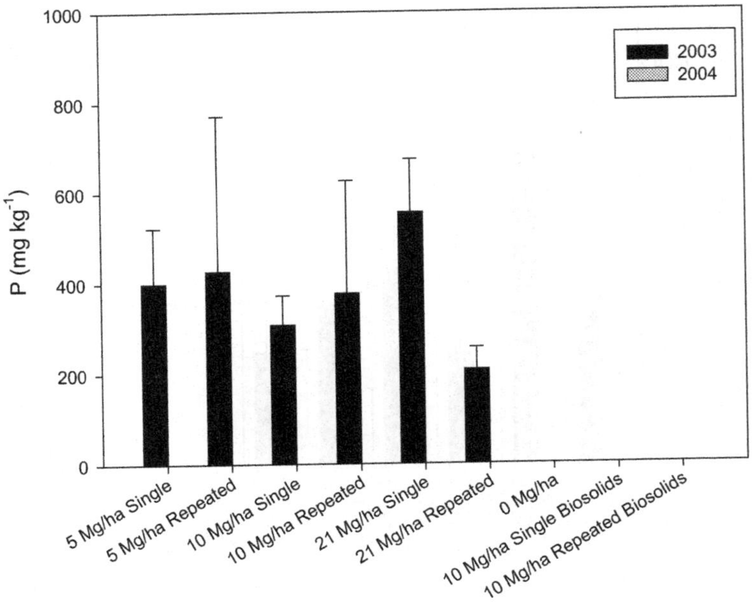

Figure 3.21 Fulvic acid P from plots of 10 Mg ha^{-1} biosolids co-applied with 5, 10, or 21 Mg ha^{-1} WTR, 0-5-cm depth, 2003 and 2004. Single and repeated co-applications occurred in 1991 and 2002, respectively. Control soils, measured only in 2004, included an untreated control and a single and repeated biosolids application. Error bars represent one standard error of the mean (n=4).

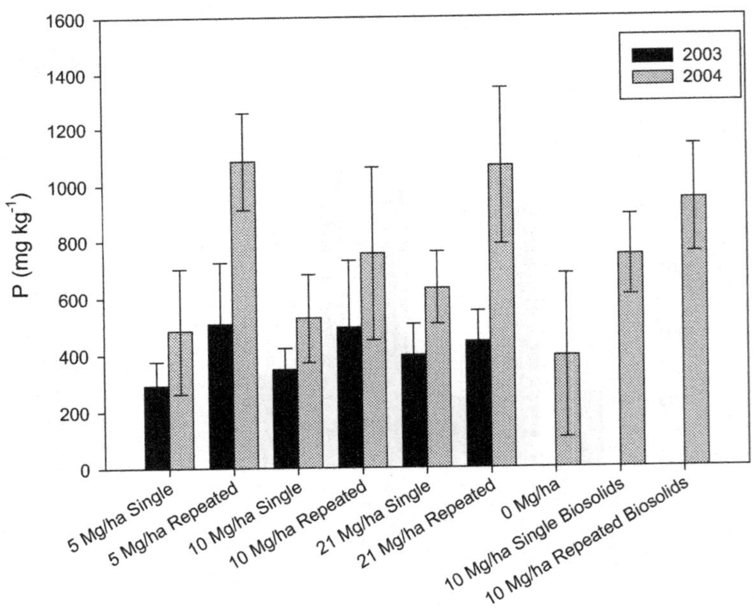

Figure 3.22 Humic acid P from plots of 10 Mg ha^{-1} biosolids co-applied with 5, 10, or 21 Mg ha^{-1} WTR, 0-5-cm depth, 2003 and 2004. Single and repeated co-applications occurred in 1991 and 2002, respectively. Control soils, measured only in 2004, included an untreated control and a single and repeated biosolids application. Error bars represent one standard error of the mean (n=4).

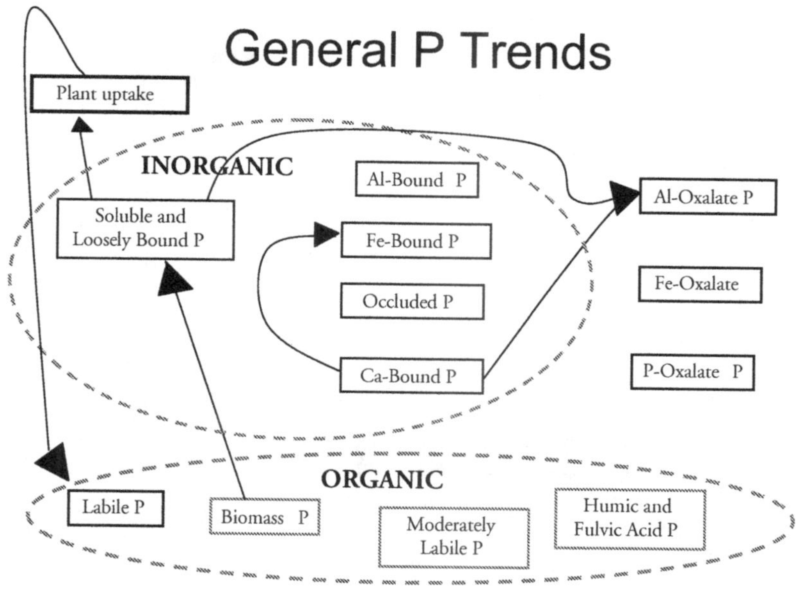

Figure 3.23 Shifts in the P pools under biosolids-WTR co-application conditions as determined by Pearson correlation matrices.

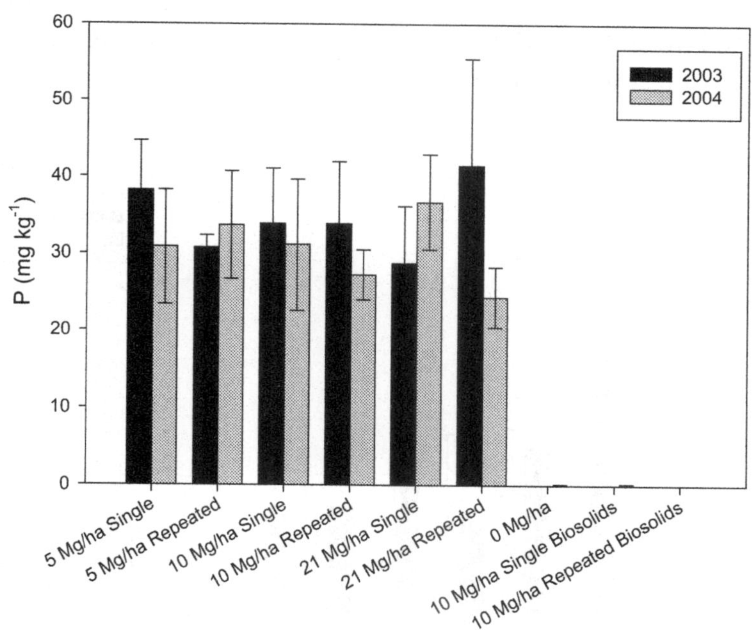

Figure 3.24 Phosphomonoesterase enzyme-cleaved P from plots of 10 Mg ha^{-1} biosolids co-applied with 5, 10, or 21 Mg ha^{-1} WTR, 0-5-cm depth, 2003 and 2004. Single and repeated co-applications occurred in 1991 and 2002, respectively. Control soils, measured only in 2004, included an untreated control and a single and repeated biosolids application. Error bars represent one standard error of the mean (n=4).

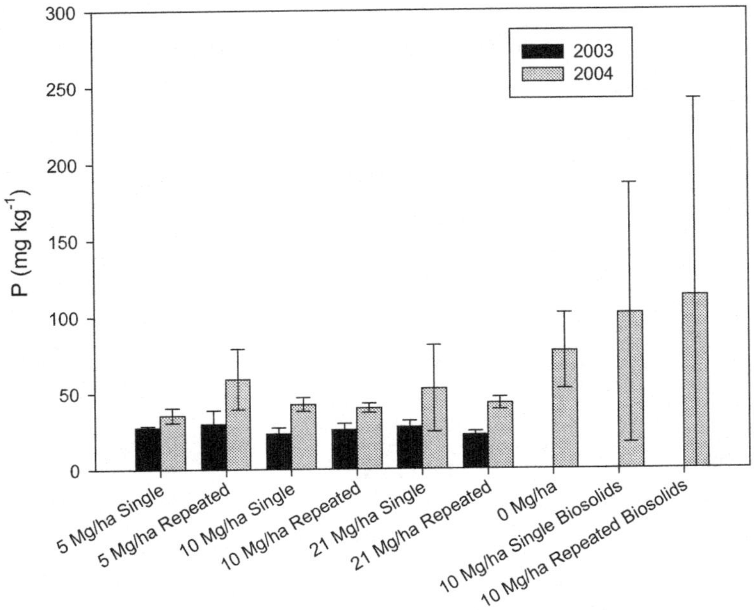

Figure 3.25 Phosphodiesterase enzyme-cleaved P from plots of 10 Mg ha^{-1} biosolids co-applied with 5, 10, or 21 Mg ha^{-1} WTR, 0-5-cm depth, 2003 and 2004. Single and repeated co-applications occurred in 1991 and 2002, respectively. Control soils, measured only in 2004, included an untreated control and a single and repeated biosolids application. Error bars represent one standard error of the mean (n=4).

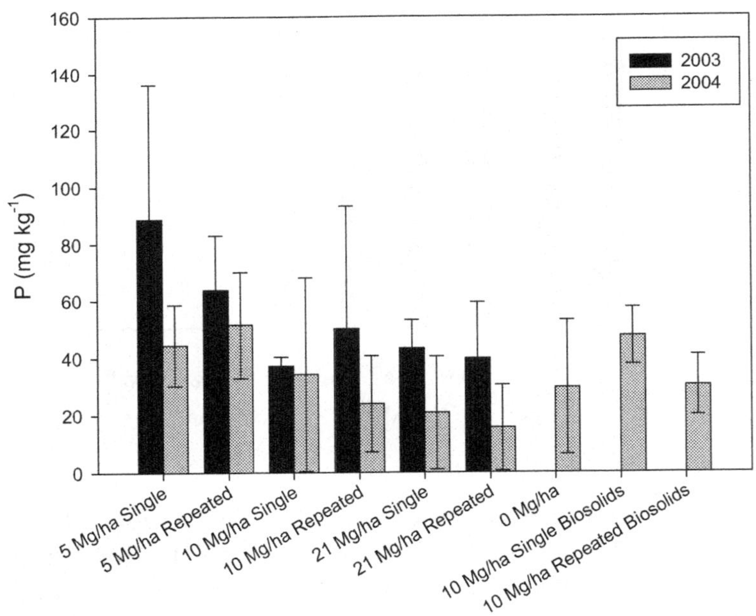

Figure 3.26 Pyrophosphatase enzyme-cleaved P from plots of 10 Mg ha^{-1} biosolids co-applied with 5, 10, or 21 Mg ha^{-1} WTR, 0-5-cm depth, 2003 and 2004. Single and repeated co-applications occurred in 1991 and 2002, respectively. Control soils, measured only in 2004, included an untreated control and a single and repeated biosolids application. Error bars represent one standard error of the mean (n=4).

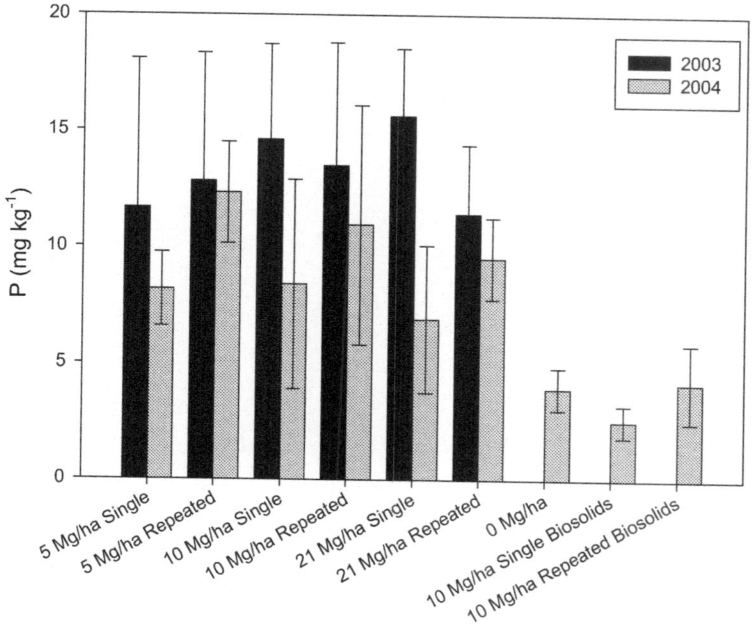

Figure 3.27 Phytase enzyme-cleaved P from plots of 10 Mg ha^{-1} biosolids co-applied with 5, 10, or 21 Mg ha^{-1} WTR, 0-5-cm depth, 2003 and 2004. Single and repeated co-applications occurred in 1991 and 2002, respectively. Control soils, measured only in 2004, included an untreated control and a single and repeated biosolids application. Error bars represent one standard error of the mean (n=4).

CHAPTER 4
SUMMARY AND CONCLUSIONS

The study objective was to quantify both the long-term effects of a single co-application of biosolids and WTR and the short-term impacts of a repeated co-application on soil P dynamics. To increase our understanding of soil P dynamics, the dominance of inorganic and organic P pools, and shifts between these pools over time as identified by the use of enzyme assays, were determined. Co-application has not been extensively studied and this is the first study to examine specific P pools under co-application conditions. In acceptance of our hypothesis, we observed similar shifts in P pools in single and repeated co-applications but observed greater P concentrations in soils that received repeated co-applications. We rejected our hypothesis that P bound in the amorphous Al fraction dominated all inorganic fractions. Although the 2004 results clearly showed this to be the case, the 2003 results did not support this contention.

General soil analysis confirmed that the addition of WTR and biosolids was responsible for increased organic C and total N. Nitrogen also increased as a result of WTR additions, although neither organic C nor total N varied directly with WTR treatment application rates. The lack of WTR effect on C and N indicated that biosolids were the major contributor of both organic C and total N.

Soil pH did not change as a result of co-application. The addition of WTR and biosolids created an influx of soluble salts in the upper soil profile as measured by EC. These salts were most likely leached below the sampling depth (0 to 5 cm) and did not reach concentrations considered hazardous to native range plants. All single application treatments had EC levels that were similar to the level in the untreated control soil. Repeated applications showed significantly greater EC values, but decreased significantly over the sampling period, showing that salts are leached over time.

Based on the Colorado P Risk Assessment, soils were analyzed for Olsen, Mehlich-III, and Bray-I P extractable concentrations. The Olsen test provided the least reliable results because it is better suited for soils with a higher pH than the study site. All three tests resulted in P concentrations considered high or very high risk for off-site P movement. According to the Risk Assessment worksheet, however, the research site would not require an actual assessment because it is not in close proximity to surface water. Although test concentrations were elevated, other researchers indicated that P sorbed to amorphous Al is stable in aqueous environments, which could mitigate some risk. In addition, if water bodies existed in close proximity, other researchers have suggested using a water extract to best predict P movement from WTR-amended soils.

Over time there was a decreasing amount of P bound to the inorganic fraction. This study found that the soluble and loosely bound P fraction was very dynamic over time. Results indicated that this fraction was receiving P from the organic biomass fraction and contributing P to the amorphous Al fraction and possibly plant uptake. Calcium-bound P was also dynamic, showing changes over time. Calcium phosphates were dissolving due to weathering, releasing P, and transferring P to the Fe-bound and the amorphous Al fractions. Even though there were P-gains in the Fe-bound fraction, the gains were relatively minute. The Fe-bound fraction, along with the Al-bound and occluded inorganic fractions, were fairly stable from 2003 to 2004, but slowly decreased over the long-term.

The oxalate extraction yielded both amorphous Al and Fe concentrations, the P associated with those particles, and the basis for calculating phosphorus saturation index (PSI). The PSI is measured on a scale of 0 to 1 with 1 meaning a soil is 100% saturated. Soils that received only biosolids had significantly greater PSI values than soils that received WTR-biosolids. The single and repeated additions of 21 Mg ha^{-1} of WTR led to a PSI reduction equal to or below the background soil PSI. Bringing PSI values below those naturally found in soils may indicate a trend toward P deficiency induced by WTR.

With the addition of WTR we thought we would observe >50 % of the P sorbed to the amorphous Al and Fe fraction. While increases were evident, >50% of the total soil P in this fraction was not observed. The most dramatic increases were observed in the 2004 Al-oxalate fraction, providing a large amount of exchange sites which easily sorb free (i.e. soluble) P, and supported our original hypothesis. Phosphorus from the soluble and loosely bound and the Ca-bound phases was being released and taken up by plants or adsorbed by amorphous Al. It was also possible that P from the organic biomass fraction released P indirectly to the amorphous Al fraction by first releasing it to the soluble and loosely bound fraction. Overall, the oxalate extractable fraction showed a significant increase from 2003 to 2004, possibly a result of weathering which created more WTR surface area.

While the oxalate extractable fraction accounted for a large amount of P, the organic fraction contained the greatest P concentration. Overall, the organic P fraction increased while the inorganic fraction decreased from 2003 to 2004, indicating an overall net immobilization effect. The increase in organic P was probably due to increased plant matter inputs resulting from increased precipitation in 2004.

Although there was a net immobilization effect, some mineralization occurred as measured by enzyme activity. Enzymes were responsible for cleaving various P forms from organic sources and shifting them to the inorganic phase. Overall, enzyme assays were unreliable for use as an accurate method for determining organic P concentrations or shifts between the P pools. The phytase enzyme may be of use for organic P fractionation prediction with further research.

While the enzymes were not reliable for P prediction, phosphomonoesterase, phytase, and phosphodiesterase showed dramatic responses to co-application. Phosphodiesterase enzyme activity in the biosolids applied alone and untreated control soils were greater than concentrations observed following WTR additions. Contradictory, phosphomonoesterase and phytase activity increased dramatically with WTR addition. The WTR addition may have created a P deficiency-like system, causing plants to respond and release these enzymes into the soil environment. Previous studies have shown that plants can exude phosphomonoesterase and phytase under P deficient conditions. The WTR treatments in this study did not produce observable plant P deficiency symptoms, but it was possible that WTR caused a false reaction and thus caused plants to secrete phosphomonoesterase and phytase.

Overall, this study was successful in determining dominant P pools and P movement between these pools over time. There were no indications that the biosolids-WTR co-application treatments caused P overloading or P deficiencies, the two major concerns of biosolids or WTR application separately. Phosphorus sorption to amorphous Al as a result of WTR application was observed, which should provide long term sorption stability. Future studies should examine other aspects of co-application including the safety of its use pertaining to human health. In regards to P dynamics, biosolids-WTR co-application is a viable option for municipal nutrient recycling and reduction of re-usable materials entering landfills.

CHAPTER 5
RECOMMENDATIONS TO UTILITIES

1. This study was successful in determining dominant P pools and P movement between these pools over time. There were no indications that a single (1991) or repeated (2002) biosolids-WTR co-application treatments caused short- or long-term P overloading or P deficiencies at the application rates used (5, 10, or 21 Mg WTR ha^{-1} co-applied with 5 Mg biosolids ha^{-1}), the two major concerns of biosolids or WTR application separately. Phosphorus sorption to amorphous Al as a result of WTR treatments will provide long-term P sorption stability. In regards to P dynamics, biosolids-WTR co-application is a viable option for municipal nutrient recycling and reduction of re-usable materials entering landfills. The addition of WTR to these systems can prolong their life expectancy with regards to long-term biosolids application and P-loading.
2. In the future, if municipalities are required to apply biosolids based on plant P requirements, not N requirements, co-application with WTR could be considered a best management practice. The co-application of 21 Mg WTR ha^{-1} + 10 Mg biosolids ha^{-1} led to phosphorus saturation index (PSI) reductions equal to or below background soil. If municipalities were required to co-apply these materials as to not drastically alter a system's ability to adsorb P, a 2:1 ratio of WTR to biosolids appears to be the optimal co-application ratio.
3. In the US, WTR are not generated to the extent of biosolids. Thus, WTR application alone or co-application with high-bearing P materials such as biosolids, can be used as a best management tool in areas where soil P runoff or leaching risk is elevated.

REFERENCES

Anderson, G. 1980. Assessing Organic Phosphorus in Soils. In *The Role of Phosphorus in Agriculture*. Edited by F.E. Khasawneh, E.C. Sample, and E.J. Kamprath. Madison, Wis.: American Society of Agronomy.

Applegate, T.J., and R. Angel. 2004. Phytase: Basics of Enzyme Function. Purdue University Cooperative Extension Publication AS-560-W [Online]. Available: <http://www.ces.purdue.edu/extmedia/AS/AS-560-W.pdf>. (cited July 10, 2006).

Barbarick, K.A., R.N. Lerch, J.M. Utschig, D.G. Westfall, R.H. Follett, J.A. Ippolito, R. Jepson, and T.M. McBride. 1992. Eight Years of Application of Biosolids to Dryland Winter Wheat. Colorado Agricultural Experiment Station Technical Bulletin TB92-1. Fort Collins, Colo.: Colorado Agricultural Experiment Station.

Basta, N.T., R.J. Zupancic, and E.A. Dayton. 2000. Evaluating Soil Tests to Predict Bermuda Grass Growth in Drinking Water Treatment Residuals With Phosphorus Fertilizer. *Jour. Environ. Qual.*, 29:2007–2012.

Bugbee, G.J., and C.R. Frink. 1985. Alum Sludge as a Soil Amendment: Effects on Soil Properties and Plant Growth. Connecticut Agricultural Experimental Station Bulletin 823. New Haven, Conn.: Connecticut Agricultural Experimental Station.

CDPHE (Colorado Department of Public Health and Environment). 2004. 5CCR 1003-7, State Board of Health Regulations Pertaining to the Beneficial Use of Water Treatment Sludge and Fees Applicable to the Beneficial Use of Sludges [Online]. Available: <http://www.cdphe.state.co.us/op/regs/waterqualitycontroldivision/100307beneficialuseofwatertreatmentsludge.pdf>. (cited November 4, 2005).

CDPHE (Colorado Department of Public Health and Environment). 1996. Revised Biosolids Regulation 4.9.0. Denver, Colo.: CDPHE.

CDPHE (Colorado Department of Public Health and Environment). 1994. Beneficial use of water treatment sludge and fees applicable to the beneficial use of sludges. 5 CCR 1003-7. Denver, Colo.: CDPHE.

City of Fort Collins (Colo.). 2006. Wastewater treatment process online tour. Available: <http://www.ci.fort-collins.co.us/wastewater/tour.php>. (cited April 26, 2006).

City of Fort Collins (Colo.). 2004. Fort Collins Utilities Drinking Water Quality Report [Online]. Available: <http://www.ci.fort-collins.co.us/water/pdf/dwqr-2004.pdf#search='Fort%20Collins%20drinking%20water'>. (cited April 26, 2006).

Cornell, D.A., and G.P. Westerhoff. 1981. Management of Water Treatment Plant Sludges. In *Sludge and Its Ultimate Disposal*. Edited by J.A. Borchardt. Ann Arbor, Mich.: Ann Arbor Science.

Correll, D.L. 1998. The Role of Phosphorus in the Eutrophication of Receiving Waters: A Review. *Jour. Environ. Qual.*, 27:261–266.

Dayton, E. A., and N.T. Basta. 2005. A Method for Determining the Phosphorus Sorption Capacity and Amorphous Aluminum of Aluminum-based Drinking Water Treatment Residuals. *Jour. Environ. Qual.*, 34:1112–1118.

Elliott, H.A., B.A. Dempsey, D.W. Hamilton, and J.R. DeWolfe. 1990. Land Application of Water Treatment Sludges: Impact and Management. Denver, Colo.: Awwa Research Foundation.

Elliott, H.A., G.A. O'Connor, P. Lu, and S. Brinton. 2002. Influence of Water Treatment Residuals on Phosphorus Solubility and Leaching. *Jour. Environ. Qual.*, 31:1362–1369.

Engelen, A. J., F.C. van der Heeft, P.H.G. Randsdorp, W.A.C. Somers, J. Schaefer, and B.J.C. van der Vat. 2001. Determination of Phytase Activity in Feed by a Colorimetric Enzymatic Method: Collaborative Interlaboratory Study. *Jour. AOAC Int.*, 84:629–633.

Harris-Pierce, R., K.A. Barbarick, and E.F. Redente. 1993. Annual Report to the City of Fort Collins, Colo.: The Effect of Sewage Sludge Application on Native Rangeland Soils and Vegetation, Fort Collins Meadow Springs Ranch. Fort Collins, Colo.: Colorado State University.

Heil, D.M., and K.A. Barbarick. 1989. Water Treatment Sludge Influence on the Growth of Sorghum-Sudangrass. *Jour. Environ. Qual.*, 18:292–298.

Ippolito, J.A., and K.A. Barbarick. 2004. Application of Anaerobically Digested Biosolids to Dryland Winter Wheat: 2002–2003 Annual Report for the Cities of Littleton and Englewood, Colo. Fort Collins, Colo.: Colorado State University.

Ippolito, J.A., and K.A. Barbarick. 2006. Phosphorus Extraction Methods for Water Treatment Residual-Amended Soils. *Comm. Soil Sci. Plant Anal.*, 37:859–870.

Ippolito, J.A., K.A. Barbarick, D.M. Heil, J.P. Chandler, and E.F. Redente. 2003. Phosphorus Retention Mechanisms of a Water Treatment Residual. *Jour. Environ. Qual.*, 32:1857–1864.

Ippolito, J.A., K.A. Barbarick, and E.F. Redente. 2002. Combinations of Water Treatment Residuals and Biosolids Affect Two Range Grasses. *Comm. Soil Sci. Plant Anal.*, 33:831–844.

Kuo, S. 1996. Phosphorus. In *Methods of Soil Analysis—Part 3: Chemical Methods*. Edited by D.L. Sparks, A.L. Page, P.A. Helmke, R.H. Loeppert, P.N. Soltanpour, M.A. Tabatabai, C.T. Johnston, and M.E. Sumner. Madison, Wis.: Soil Science Society of America.

Loeppert, R.H., and W.P. Inskeep. 1996. Iron. In *Methods of Soil Analysis—Part 3: Chemical Methods*. D.L. Sparks, A.L. Page, P.A. Helmke, R.H. Loeppert, P.N. Soltanpour, M.A. Tabatabai, C.T. Johnston, and M.E. Sumner. Madison, Wis.: Soil Science Society of America.

Lucas, J.B., T.A. Dillaha, R.B. Reneau, J.T. Novak, and W.R. Knocke. 1994. Alum Sludge Land Application and Its Effect on Plant Growth. *Jour. AWWA*, 86:75–83.

Maguire, R.O., J.T. Sims, and F.J. Coale. 2000. Phosphorus Fractionation in Biosolids-Amended Soils: Relationship to Soluble and Desorbable Phosphorus. *Soil Sci. Soc. Am. Jour.*, 64:2018–2024.

Makris, K.C., W.G. Harris, G.A. O'Connor, and T.A. Obreza. 2004. Phosphorus Immobilization in Micropores of Drinking Water Treatment Residuals: Implications for Long-term Stability. *Environ. Sci. Technol.*, 38:6590–6596.

Makris, K.C., W.G. Harris, G.A. O'Connor, T.A. Obreza, and H.A. Elliott. 2005. Physicochemical Properties Related to Long-term Phosphorus Retention by Drinking Water Treatment Residuals. *Environ. Sci. Technol.*, 39:4280–4289.

Mehlich, A. 1984. Mehlich 3 Soil Test Extractant: A Modification of Mehlich 2 Extractant. *Comm. Soil Sci. Plant Jour.*, 15:1409–1416.

Miller, S.S., J. Liu, D.L. Allan, C.J. Menzhuber, M. Fedorova, and C.P. Vance. 2001. Molecular Control of Acid Phosphatase Secretion Into the Rhizosphere of Proteoid Roots From Phosphorus-Stressed White Lupin. *Plant Physiol.*, 127:594–606.

Mulvaney, R.L. 1996. Nitrogen–Inorganic Forms. In *Methods of Soil Analysis—Part 3: Chemical Methods*. Edited by D.L. Sparks, A.L. Page, P.A. Helmke, R.H. Loeppert, P.N. Soltanpour, M.A. Tabatabai, C.T. Johnston, and M.E. Sumner. Madison, Wis.: Soil Science Society of America.

Nelson, D.W., and L.E. Sommers. 1996. Total Carbon, Organic Carbon, and Organic Matter. In *Methods of Soil Analysis—Part 3: Chemical Methods*. Edited by D.L. Sparks, A.L. Page, P.A. Helmke, R.H. Loeppert, P.N. Soltanpour, M.A. Tabatabai, C.T. Johnston, and M.E. Sumner. Madison, Wis.: Soil Science Society of America.

Novak, J.M., and D.W. Watts. 2004. Increasing the Phosphorus Sorption Capacity of Southeastern Coastal Plain Soils Using Water Treatment Residuals. *Soil Sci.*, 3:206–214.

NOAA (National Oceanic and Atmospheric Administration). 2003–2004. Annual climatological summary, station 053006/99999 [Online]. Available: <http://ncdc.noaa.gov/onlinestore.html>. (cited December 1, 2005).

NRCS (National Resources Conservation Service). 1980. Soil Survey of Larimer County Area, Colorado [Online]. Available: <http://soils.usda.gov/survey/online_surveys/colorado/larimer/Text-Part%201.pdf>. (cited November 4, 2005).

Rengasamy, P., J.M. Oades, and T.W. Hancock. 1980. Improvement of Soil Structure and Plant Growth by Addition of Alum Sludge. *Comm. Soil Sci. Plant Anal*,. 11:533–545.

Rhoades, J.D. 1996. Salinity: Electrical Conductivity and Total Dissolved Solids. In *Methods of Soil Analysis—Part 3: Chemical Methods*. Edited by D.L. Sparks, A.L. Page, P.A. Helmke, R.H. Loeppert, P.N. Soltanpour, M.A. Tabatabai, C.T. Johnston, and M.E. Sumner. Madison, Wis.: Soil Science Society of America.

Rodriguez, J.B., J.R. Self, and P.N. Soltanpour. 1994. Optimal Conditions for Phosphorus Analysis by the Ascorbic Acid–Molybdenum Blue Method. *Soil Sci. Soc. Am. Jour.*, 58:866–870.

SAS Institute. 2002. *SAS/STAT User's Guide*. Version 9.1. Cary, N.C.: SAS Institute.

Sharkoff, J.L., R.M. Waskom, and J.G. Davis. 2003. Colorado Phosphorus Index Risk Assessment. Agronomy Technical Note No. 95, U.S. Dept. of Agriculture (USDA), NRCS. Lakewood, Colo.: USDA-NRCS.

Sherrod, L.A., G. Dunn, G.A. Peterson, and R.L. Kolberg. 2002. Inorganic Carbon Analysis by Modified Pressure-Calcimeter Method. *Soil Sci. Soc. Amer. Jour.*, 66:299–305.

Soltanpour, P.N., G.W. Johnson, S.M. Workman, J.B. Jones Jr., and R.O. Miller. 1996. Inductively Coupled Plasma Emission Spectrometry and Inductively Coupled Plasma–Mass Spectrometry. In *Methods of Soil Analysis—Part 3: Chemical Methods*. Edited by D.L. Sparks, A.L. Page, P.A. Helmke, R.H. Loeppert, P.N. Soltanpour, M.A. Tabatabai, C.T. Johnston, and M.E. Sumner. Madison, Wis.: Soil Science Society of America.

Sui, Y., M.L. Thompson, and C. Shang. 1999. Fractionation of Phosphorus in a Mollisol Amended With Biosolids. *Soil Sci. Soc. Am. Jour.*, 63:1174–1180

Tabatabai, M.A. 1994. Soil Enzymes. In *Methods of Soil Analysis—Part 2: Microbiological and Biochemical Properties*. Edited by R.W. Weaver, J.S. Angle, and P.S. Bottomley. Madison, Wis.: Soil Science Society of America.

Thomas, G.W. 1996. Soil pH and Soil Acidity. In *Methods of Soil Analysis—Part 3: Chemical Methods*. Edited by D.L. Sparks, A.L. Page, P.A. Helmke, R.H. Loeppert, P.N. Soltanpour, M.A. Tabatabai, C.T. Johnston, and M.E. Sumner. Madison, Wis.: Soil Science Society of America.

USEPA (U.S. Environmental Protection Agency). 1999. Biosolids Generation, Use and Disposal in the United States [Online]. Available: <http://www.ees.ufl.edu/homepp/townsend/Courses/SHWM/F03/ARM/biosolids.pdf>. (cited November 4, 2005).

Weaver, R.M. 1974. A Simplified Determination of Reductant-Soluble Phosphate in Soil Phosphate Fractionation Schemes. *Soil Sci. Soc. Amer. Proc.*, 38:153–154.

Zhang, H., and J.L. Kovar. 2002. Phosphorus fractionation. In *Methods of Phosphorus Analysis for Soils, Sediments, Residuals, and Waters*. Southern Coop. Series Bull. #396. Edited by G.M. Pierzynski. Raleigh, N.C.: North Carolina State University.

ABBREVIATIONS

ANOVA	analysis of variance
EC	electrical conductivity
ICP-AES	inductively coupled plasma – atomic emission spectrometry
LSD	least significant difference
P	probability level
PSI	phosphorus saturation index
WTR	water treatment residuals

IWA Publishing's authorised EU representative for General Product Safety Regulations is Diane D'Arras, 15 rue Duret, 75116 Paris, France, e-mail: safety@iwap.co.uk.

Printed and bound by CPI Group (UK) Ltd, Croydon, CR0 4YY
17/09/2026
02218804-0010